THE GHOSTS OF STALINGRAD

A thesis presented to the Faculty of the U.S. Army
Command and General Staff College in partial
fulfillment of the requirements for the
degree

MASTER OF MILITARY ART AND SCIENCE
Military History

by

WILLARD B. AKINS II, MAJ, USAF
B.S., Unites States Air Force Academy, Colorado Springs, Colorado, 1989

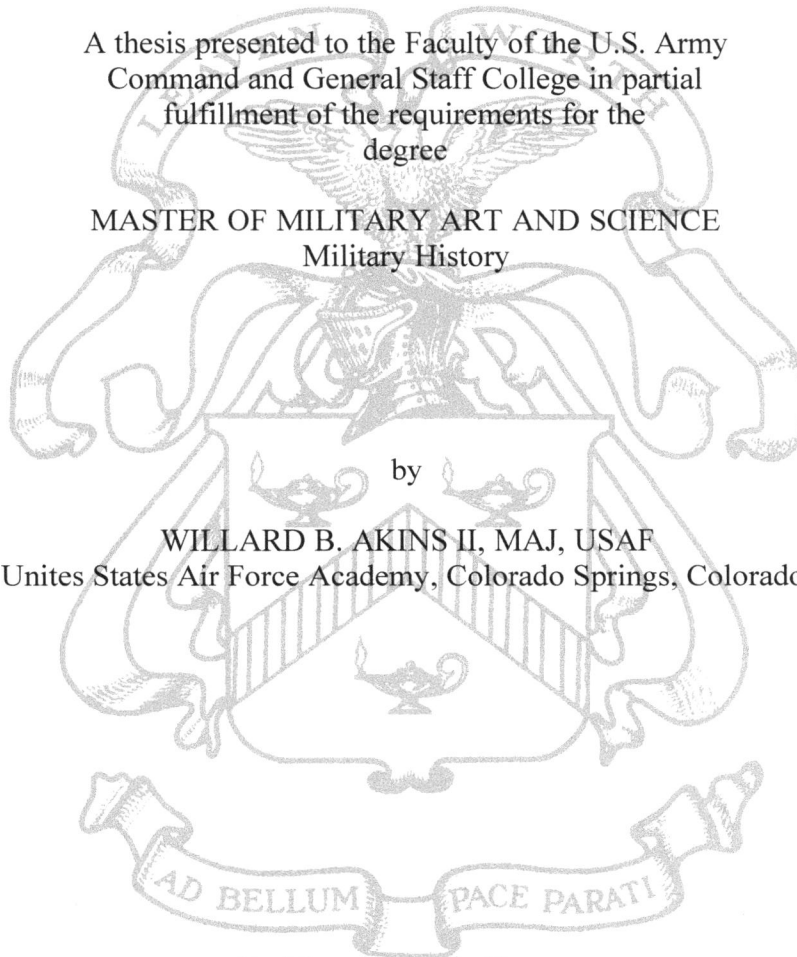

Fort Leavenworth, Kansas
2004

MASTER OF MILITARY ART AND SCIENCE

THESIS APPROVAL PAGE

Name of Candidate: Major Willard B. Akins II

Thesis Title: The Ghosts of Stalingrad

Approved by:

_____, Thesis Committee Chair
Lieutenant Colonel John A. Suprin, M.A.

_____, Member
Samuel J. Lewis, Ph.D.

_____, Member
Colonel William J. Heinen, M.M.A.S.

Accepted this 18th day of June 2004 by:

_____, Director, Graduate Degree Programs
Robert F. Baumann, Ph.D.

The opinions and conclusions expressed herein are those of the student author and do not necessarily represent the views of the U.S. Army Command and General Staff College or any other governmental agency. (References to this study should include the foregoing statement.)

Report Documentation Page

1. REPORT DATE **17 JUN 2004**	2. REPORT TYPE	3. DATES COVERED **-**

4. TITLE AND SUBTITLE **Ghosts of Stalingrad**	5a. CONTRACT NUMBER
	5b. GRANT NUMBER
	5c. PROGRAM ELEMENT NUMBER
6. AUTHOR(S) **Williard Akins**	5d. PROJECT NUMBER
	5e. TASK NUMBER
	5f. WORK UNIT NUMBER

7. PERFORMING ORGANIZATION NAME(S) AND ADDRESS(ES) **US Army Command and General Staff College,1 Reynolds Ave,Fort Leavenworth,KS,66027-1352**	8. PERFORMING ORGANIZATION REPORT NUMBER **ATZL-SWD-GD**

9. SPONSORING/MONITORING AGENCY NAME(S) AND ADDRESS(ES)	10. SPONSOR/MONITOR'S ACRONYM(S)
	11. SPONSOR/MONITOR'S REPORT NUMBER(S)

12. DISTRIBUTION/AVAILABILITY STATEMENT
Approved for public release; distribution unlimited

13. SUPPLEMENTARY NOTES

14. ABSTRACT

The Battle of Stalingrad was a disaster. The German Sixth Army consisted of over 300,000 men when it approached Stalingrad in August 1942. On 2 February 1943, 91,000 remained; only some 5,000 survived Soviet captivity. Largely due to the success of previous aerial resupply operations, Luftwaffe leaders assured Hitler they could successfully supply the Sixth Army after it was trapped. However, the Luftwaffe was not up to the challenge. The primary reason was the weather, but organizational and structural flaws, as well as enemy actions, also contributed to their failure. This thesis will address why the Demyansk and Kholm airlifts convinced the Germans that airlift was a panacea for encircled forces; the lessons learned from these airlifts and how they were applied at Stalingrad; why Hitler ordered the Stalingrad airlift despite the logistical impossibility; and seek out lessons for todayŸs military. The primary reason for the Stalingrad tragedy was that GermanyŸs strategic leadership did not apply lessons learned from earlier airlifts to the Stalingrad airlift, and the U.S. military is making similar mistakes with respect to the way it is handling its lessons learned from recent military operations

15. SUBJECT TERMS

16. SECURITY CLASSIFICATION OF:			17. LIMITATION OF ABSTRACT **1**	18. NUMBER OF PAGES **105**	19a. NAME OF RESPONSIBLE PERSON
a. REPORT **unclassified**	b. ABSTRACT **unclassified**	c. THIS PAGE **unclassified**			

ABSTRACT

THE GHOSTS OF STALINGRAD, by Major Willard B. Akins II, 88 pages.

The Battle of Stalingrad was a disaster. The German Sixth Army consisted of over 300,000 men when it approached Stalingrad in August 1942. On 2 February 1943, 91,000 remained; only some 5,000 survived Soviet captivity. Largely due to the success of previous aerial resupply operations, Luftwaffe leaders assured Hitler they could successfully supply the Sixth Army after it was trapped. However, the Luftwaffe was not up to the challenge. The primary reason was the weather, but organizational and structural flaws, as well as enemy actions, also contributed to their failure.

This thesis will address why the Demyansk and Kholm airlifts convinced the Germans that airlift was a panacea for encircled forces; the lessons learned from these airlifts and how they were applied at Stalingrad; why Hitler ordered the Stalingrad airlift despite the logistical impossibility; and seek out lessons for today's military. The primary reason for the Stalingrad tragedy was that Germany's strategic leadership did not apply lessons learned from earlier airlifts to the Stalingrad airlift, and the U.S. military is making similar mistakes with respect to the way it is handling its lessons learned from recent military operations.

TABLE OF CONTENTS

Page

MASTER OF MILITARY ART AND SCIENCE THESIS APPROVAL PAGE ii

ABSTRACT iii

FIGURES v

THESIS OBJECTIVES 1

CHAPTER 1. INTRODUCTION TO THE LUFTWAFFE AND AERIAL RESUPPLY .. 3

CHAPTER 2. DANGEROUS PRECEDENTS: DEMYANSK AND KHOLM 17

Demyansk 18
Kholm 27

CHAPTER 3. STALINGRAD 31

CHAPTER 4. STRATEGIC DILETTANTISM 61

CHAPTER 5. CONCLUSION 81

Relevance to the Contemporary Environment 83

FIGURES 91

SELECTED BIBLIOGRAPHY 93

Books 93
U.S. Government Sponsored Documents 95
Primary 95
Secondary 95
Dissertations and Theses 96
Periodicals and Newspaper Articles 97

INITIAL DISTRIBUTION LIST 98

CERTIFICATION FOR MMAS DISTRIBUTION STATEMENT 99

FIGURES

Page

Figure 1. The Demyansk and Kholm Pockets 91

Figure 2. Luftwaffe Air Supply Corridor 92

THESIS OBJECTIVES

Problem Statement: The Stalingrad airlift was a failure. Its breakdown presented disastrous consequences for the Sixth Army. Certainly "General Winter" played a key role in the overall catastrophe, but the Germans had already suffered through the Russian winter of 1941 and should have know what was in store for them at Stalingrad just one short year later. This research will investigate the Stalingrad airlift, search for parallels that may exist in today's political-military environment, and present lessons that are applicable in current military operations.

Research Question: What factors contributed to the failure of the aerial resupply efforts at Stalingrad? Subordinate questions: How did leadership affect the outcome? How were transport units organized? What was their doctrine and how was it applied? What aircraft were used? What lessons did the Soviets learn from the Kholm/Demyansk airlifts and how did they apply them to thwart the Stalingrad efforts; how successful were they? Are there any lessons we can apply to United States military operations?

Thesis Statement: The purpose of this paper is to explain the failure of the resupply at Stalingrad, and to look for parallels in the current US civil-military dynamic, suggesting an inexorable march along the same path to failure.

Qualifications: I am motivated to conduct this research project. I will use this research to complete an historical Master of Military Art and Science (MMAS) degree. I have had an interest in the Luftwaffe and Russia since I was young. I have been highly interested in Operation BARBAROSSA and the Battle of Stalingrad since I learned about it in some detail while taking Air Command and Staff College by correspondence two years ago. I am a USAF Weapons Officer and C-130 pilot with 3,500 hours. I am an Air

1

Force combined-arms expert and have a thorough understanding of airpower, although my particular specialty is tactical airlift.

Limitations: There are several limitations to this study of the Luftwaffe and its impact on the Stalingrad airlift. First, few primary sources exist, since many of the Luftwaffe's records were destroyed immediately following the war. Hitler, Göring, Milch, and others are not available for interview. Second, some secondary sources written after the war by former Luftwaffe officers and Stalingrad participants are considered tainted if not grossly biased. Third, translations also run the risk of being biased, through either malicious intent or pure accident, and there is always the possibility that certain words and concepts will not translate completely or coherently into other languages, forcing the translator to ad lib. Fourth, I do not speak or read German or Russian with enough ability to be of any use in this project. Finally, I will not have the opportunity to visit the countries, locations, or battlefields mentioned and discussed in the text.

CHAPTER 1

INTRODUCTION TO THE LUFTWAFFE AND AERIAL RESUPPLY

> I have done my best, in the past few years, to make our Luftwaffe the largest and most powerful in the world. The creation of the Greater German Reich has been made possible largely by the strength and constant readiness of the Air Force. Born of the spirit of the German airmen in the First World War, inspired by faith in our Fuehrer and Commander-in-Chief—thus stands the German Air Force today, ready to carry our every command of the Fuehrer with lightening speed and undreamed-of might.[1]

> Reichsmarschall Hermann Göring, *The Rise and Fall of the German Air Force: 1933 to 1945*

Before the onset of World War II, Germany faced the daunting prospect of building an air force that, owing to the restrictions placed upon it by the Treaty of Versailles, was in an embryonic state. Having been forced to make concessions at the end of World War I that limited the size of its army and navy and prohibited it from maintaining an active military airpower, Germany was not in an enviable position. General Hans von Seeckt argued vehemently that Germany should be allowed to maintain an independent air force of 1,700 aircraft and 10,000 men. He was unsuccessful. Nevertheless, due to his foresight and vigilance, Germany succeeded in securing a cadre of experienced aviation officers who remained hidden within the staffs of the army and navy. This cadre kept the ideas of a strong, resurgent air force resonating throughout the interwar period. Under the direction of Reichsmarschall Hermann Göring, one of Adolf Hitler's most trusted advisors, Germany was able to create a viable, aerial armada that would challenge Europe, Asia, and North Africa: the Luftwaffe. By 1940, the Luftwaffe was the biggest air force in the world.[2]

With the ascension of Hitler and the Nazi regime, Germany turned its thoughts toward mobilization and rearmament. A large portion of these efforts went into building military airpower. Because of the advances Germany had made in civilian aviation, its militarization was also within Germany's reach. Like their British counterparts, the Germans felt the future of military airpower lay in the development of an independent bomber force. However, technological limitations and doctrinal disagreements between members of the Luftwaffe general staff, as well as demands upon the German Air Force in support of the German Army, stopped the strategic bombing effort before it got off the ground.[3]

German airpower doctrine consisted of several elements, noted historian Richard Muller, making it difficult to summarize. The Luftwaffe grew to maturity in the 1930s and was philosophically centered around the long-range bomber, but design problems with German aviation-engine technology and the needs of the army dictated otherwise. German airpower theorists of the era were aware that twentieth-century industrial societies were extremely vulnerable to aerial attack. As a result, much of the debate and theorizing between the wars revolved around the use of independent or strategic airpower. German Air Force officers, no less than their European and American counterparts, were ardent proponents of an independent bomber force. The airpower doctrine that eventually emerged in Germany was a hodgepodge of joint elements, with airpower utilized to support the other services, and independent strikes intended to destroy the enemy's war economy or other centers of gravity.[4]

Germany's geographic position focused the Luftwaffe on supporting the German Army. Germany did not have the advantage of some countries in terms of geostrategic position. Britain, for example, was situated so that strategic bombing would be the only practical way she could hope militarily to influence continental Europe; Germany did not have that luxury. Surrounded on two sides by perceived, if not real enemies, Germany's focus on airpower was more geared toward combined-arms warfare and support of the German Army.[5] In fact, Hitler's foreign policy stated that only France and Poland were potential enemies.[6] With these adjacent, putative enemies, there was no need for a true long-range, strategic bomber, and no need to focus on developing a doctrine for deep, strategic attacks. If war was to break out, Germany's position within the hub of the European continent meant that she was not in any position to ignore the demands of the army. Doing so would put Germany at a distinct disadvantage. As a result, the Luftwaffe directed a majority of its energy to the army's support.[7] Since the army was by far the larger and more important of the military services, the primary task of the German Air Force would be to support its maneuvers: the Luftwaffe could achieve this by destroying enemy troop concentrations, strong points, and lines of communication.[8] This utilitarian application of airpower was the result of extensive study and institutionalization of the tactical and operational lessons of World War I.[9] In addition to bombing and other types of destructive activities, the Luftwaffe also contributed a great deal to the rapid mobility and supply of the Wehrmacht. The need for airlift to successfully execute these military operations is axiomatic.

The transport aircraft, mainly Junkers 52s and Heinkel 111s, enabled the Luftwaffe to keep pace with German Army advances, and later with its retreats. According to Asher Lee, a fleet of several hundred transport aircraft was always available to continue operations. Bombs, troops, fuel, spare parts, ground staff, and other equipment could be moved within twenty-four hours to occupy and operate from advance landing grounds, which they may have been bombing only several days before. Transport aircraft were also used for medical evacuation as well as for flying equipment, fuel, and stores to forward troops, airdropping supplies to troops when necessary, and for hastening airborne or parachute troops to vulnerable points in the line.[10]

Germany's participation in the Spanish Civil War taught them many lessons about the application of airpower.[11] From July 1936 until April 1939, Adolf Hitler assisted General Francisco Franco against the Spanish Republic in the Spanish Civil War. During this operation, twenty Junkers Ju-52s ferried 10,000 Moroccan troops, as well as the Spanish Foreign Legion and their equipment from Tetuan to Spain in a ten-day period, enabling Franco to consolidate forces and establish a firm position to launch an offensive against the government.[12] This was the world's first large-scale airlift operation.[13] Hitler remarked on this accomplishment in September 1942, "Franco ought to erect a monument to the glory of the Ju 52. It is this aircraft that the Spanish revolution has to thank for its victory."[14] It is perplexing that Hitler and the Luftwaffe never realized the advantages of a large, modern, and robust airlift fleet for the mobility and long-term success of their own military until it was too late.

One of the foremost reasons behind Hitler's willingness to support Franco's cause was to test Germany's new military equipment. This testing validated the strategy and tactics that had been developed in Germany. As a result, the Luftwaffe leadership entered the war confident that they had found both the means and the application to fulfill their objectives.[15] These discoveries greatly contributed to the fiscal resources Hitler gave to his air force.

Nevertheless, the Luftwaffe suffered from fiscal constraints as did the other services. However, there were two reasons that allowed the German Air Force to secure its necessary financing. First, Hitler had great confidence in airpower. He had seen the aircraft circling over the front lines of WWI and recognized their military value; to him the importance of possessing a strong air force was self-evident.[16] As a result, Hitler placed a strong emphasis on the Luftwaffe.[17] Second, the number two man in Hitler's Nazi regime, Hermann Göring, was the Luftwaffe's Commander-in-Chief. His position within the Nazi regime eased the struggle of obtaining the fiscal resources necessary to build a strong, efficacious air force. Göring's immediate access to Hitler, along with the latter's respect for his views on aviation matters, proved a priceless asset in the fight for the Reich's precious financial and economic resources during rearmament, an asset which neither the Defense Minister, Field Marshal Werner von Blomberg, nor the commanders-in-chief of the other two services possessed.[18] In addition to the fiscal limitations the Germans faced in the late thirties, Germany's lack of natural resources also presented dilemmas to senior officials responsible for dividing these same materials to the industries that were expanding Germany's military might. Despite Germany's scarce resources, Hitler had a strategy for how he planned to achieve his goals, and his growing military was the centerpiece.

He hoped to create for Germany an "invulnerable" position in Europe and the world no matter how long it would take; in the German plan there was little concern for the destruction of industrial and war production.[19] His strategy was based upon overwhelming Germany's foes with superior numbers and firepower, requiring huge increases in all the armed services. Interservice rivalry and competing interests for armaments created fierce competition for the Reich's resources: iron, steel, copper, tin, rubber, oil, etcetera became increasingly scarce.[20] This paucity of natural resources, combined with an approach that called for overwhelming firepower in support of the army, made it nearly impossible for long-range, strategic-bombing advocates to galvanize others to their cause. To achieve their führer's objectives, all the Wehrmacht required was the support of combined-arms aircraft; as a result, Luftwaffe production reflected this need. Medium-range bombers, ground-attack aircraft, etcetera, were designed and built to support the army.

Despite Hitler' assertions to the contrary, many German leaders never desired nor anticipated a protracted war. They envisioned quick, decisive military victories that obviated any potential necessity to destroy an enemy's industrial base. Göring himself stated after the war that he was always against the invasion of Russia even though he was confident that the Luftwaffe was easily the master of their eastern adversary: "I knew that we could defeat the Russian Army; but how were we ever to make peace with them? After all, we could not march to Vladivostock!"[21] He only acquiesced due to Hitler's insistence on the campaign. The Reichsmarschall's reticence was ostensibly due to a presumption that such a campaign would never be decisively short. The desire for quick, decisive victories led the Germans to underestimate the number of aircraft and pilots necessary to secure Germany's goals should a short campaign fail. The Luftwaffe's leadership contributed to this underestimation. In particular the Luftwaffe Chief of Staff, General Hans Jeschonnek, was stricken by the forlorn hope that a war with Russia would be brief, eliminating the need for long term planning; therefore, he possessed little interest in Luftwaffe training, concerning himself with the already available strategic-tactical force.[22]

The German Air Staff was fundamentally organized like any other air staff. Göring was at the top; Field Marshal Erhard Milch was his deputy, the Secretary of State, and the Inspector General; Jeschonnek his Chief of Staff; and General Ernst Udet was Chief of Aircraft Design and Supply. Göring had an operational staff which dealt with all major issues of policy and Luftwaffe organization, anti-aircraft defense, operations in the field, weather, intelligence, security, the German aircraft industry, etcetera. There was also a separate staff that dealt with administration and maintenance, once the operational policy was implemented. However, the Air Ministry exercised most of its control over fielded-units through a series of inspectorates, all under Milch. These were the link between fielded, operational units and the Air Ministry in Berlin, and ensured the former followed the policies of the latter.[23]

German leaders organized the Luftwaffe into geographic air fleets. There were four before the war with two more created as the war progressed to account for newly acquired territory (during German expansion). Number five was added to cover Denmark and Norway. In 1943 a sixth air fleet was created to handle the responsibilities of the Central Russian Front. Its headquarters was at Smolensk, but only briefly as the tide turned due to the Soviet army inexorably pushing the Wehrmacht back toward Germany.

Each German air fleet usually varied between five hundred and one thousand operational bomber, fighter, and reconnaissance aircraft. The air fleets' operational strength varied according to its commitments. Each air fleet was organized into two or three air divisions (or air corps depending on the year and whim of the air staff) of between 250 and 500 aircraft. The German air staff was never satisfied with the air division or air corps as an operational unit. They favored employment through smaller, tactical air commands. These smaller units were responsible for less airspace and could therefore focus wholeheartedly on the army formations they supported. Their aircraft strength fluctuated between 100 to 250 bombers, fighters and reconnaissance aircraft. Regardless of the echelon, German operational commands were designed to closely support German army operations.[24]

Conspicuously absent from the air fleet hierarchy were the air transports. This absence was to have huge repercussions as the war progressed. German air transport was given little thought, if it was given any thought at all. This is quite perplexing given the success of the Luftwaffe airlift during the Spanish Civil War. The Germans must have realized the importance the airlift made in determining the outcome in favor of the Spanish dictator. Yet air transport was largely ignored.[25]

Since the air transports were an afterthought, the German Air Force neglected establishing a proper, efficient organizational structure for them. As a result, the transports did not have an air fleet of their own; they were simply a subordinate wing of the 7th Air Division. This subservient role was a serious detriment to the air transport forces. Professor Richard Suchenwirth indicates that the establishment of an independent air transport fleet would have afforded them an air fleet commander-in-chief and staff made of general staff officers. Experienced and high-ranking officers would provide the clarity of thought to tackle administrative and other day-to-day problems. The transport air staff would automatically provide solutions to problems identified during war games. These high-ranking airlift experts, whose inputs could not be callously ignored, would have been on-hand to answer questions arising during the discussion of the first large-scale air-supply operation on the Eastern Front. Their existence may have prevented the wasteful, pointless air supply operations at Kholm and Demyansk, not to mention Stalingrad.[26] Failure to establish an independent air-transport air fleet was one of the primary reasons behind the decision to carry out the Stalingrad airlift. Organization, however, was not the only problem the transport fleet faced; they also suffered from training problems that were systemic to the entire Luftwaffe.

Prior to the war and into 1942, a German pilot received almost one hundred hours' flying time before earning his wings. Luftwaffe pilot training started with the A course, which consisted of about thirty flying hours in basic aircraft. Five hours were dual-control flying with an instructor and twenty-five were solo. Following the A course, pilots proceeded to the B course, which provided an additional sixty hours of flight time in more powerful aircraft. The B course was the discriminator. It determined whether a candidate pilot was best suited as a fighter pilot, bomber pilot, observer, or alternately, ground crew.[27]

The C course instructed bomber and long-distance reconnaissance pilots on multi-engine aircraft, mainly the Junkers 52, Dornier 17, and older Heinkel 111s. Specialized, air-transport basic training consisted of: formation flying, low-level flying, landing techniques, forced landings, night and instrument flying, and personnel and equipment airdrops. Pilots who had completed the above training were fully qualified for assignment to an air transport unit. The above training was the standard under ideal, peacetime conditions. The exigencies of war and strategic decisions of Luftwaffe leadership chipped-away at an initially efficacious training program until, by the end of the war, the

11

training program was a shell of its former self.[28] Through the late summer of 1942, German pilots received at least as many training hours as their counterparts in the Royal Air Force. By 1943, the numbers began an insidious slide to the detriment of the Germans such that by the last half of the year Luftwaffe pilots were completing barely one-half the training hours allocated to enemy pilots. This disparity was even greater in operational aircraft: one-third of the Royal Air Force total and one-fifth of the American total.[29]

Even more serious than the long-term degradation of the training program was the concomitant loss of the competitive, high-caliber personnel who became aviators. The Luftwaffe possessed the best and brightest of German youth. Under interrogation after the war, Göring stated that, ". . . the Luftwaffe had first priority and thus the cream of Germany, the U-boats were second, and the panzers third. Even at the end the best of German youth went into the Luftwaffe. . . ."[30] As the war progressed and the German Air Force began hemorrhaging at an almost exponential rate, pilot attrition forced younger and less-qualified personnel behind aircraft controls, further reducing Luftwaffe effectiveness.

The aircraft selected for the transport forces was the venerable Ju-52. It was renowned for its versatility and, due to its versatility, was the primary training aircraft in the schools. The Germans never replaced the Ju-52 with an aircraft exclusively created for pilot training.[31] As a result, there were competing interests. The Ju-52 was tied to the training program, yet this same aircraft was useful in many other applications. Until 1943, all requests for the Ju-52 came at the expense of the Chief of Training, and new production of Ju-52s never met the demands.[32]

Further convoluting the mix was that German officials had setup a command structure for the air transport forces that ignored the traditional proverb that warns against serving two masters. The Luftwaffe's highest ranking air transport officer was the commanding officer of the instrument flight school as well as the Air Transport Officer, who had command of the Air Transport Staff and air transport units. Brigadier General Fritz Morzik, who became the Air Transport Officer on 1 October 1941 as a colonel, described the tensions and difficulties of serving two masters:

> Inasmuch as the Air Transport Officer was at the same time the commanding officer of the instrument flight schools, his position was one of dual subordination. As Air Transport Officer he belonged to the staff of the Quartermaster General, and as commanding officer of the instrument flight schools to that of the Chief of Training. Inevitably a certain amount of friction resulted, which also made its effects felt among the flying units and in the schools. Requests and pleas for a clarification of the situation were without avail, and a tug-of-war within the command headquarters ensued. The functions of the air transport officer and those of the commanding officer of the instrument flight schools were united in one man who had two operational command staffs under him, one for each of his functions. This unfortunate solution to the problem of the command of two important activities was bound to result in the unintentional neglect of one of them.[33]

The end result was the attrition of a prolonged, multifront war. The primary training aircraft were constantly borrowed to sustain military operations, and the destruction of these same aircraft through enemy actions slowly bled the transport air fleet in particular, and the Luftwaffe as a whole, to death.

---

[1]British Air Ministry, *The Rise and Fall of the German Air Force: 1933 to 1945* (New York: St. Martin's Press, 1983), xvii. Göring made this statement in August 1939.

[2]David Irving. *The Rise and Fall of the Luftwaffe: The Life of Luftwaffe Marshal Erhard Milch* (Boston: Little, Brown and Company, 1973), 336.

[3]Williamson Murray, *Strategy for Defeat: The Luftwaffe, 1933-1945* (Maxwell Air Force Base: Air University Press, 1983), 3-11.

[4]Richard R. Muller, "The Luftwaffe and BARBAROSSA, 1941," In *Distance Learning Version 3.0: Military Studies*, by the Air Command and Staff College (Maxwell Air Force Base: Air Command and Staff College, 2000), 368; for a comprehensive, yet concise account of the development of German airpower doctrine between the wars, see Murray, 3-11.

[5]Willamson Murray, "Strategic Bombing: The British, American, and German experiences," in *Military Innovation in the Interwar Period,* ed. Williamson Murray and Allan R. Millett (Cambridge: Cambridge University Press, 1998), 139; Ibid, 112-113.

[6]Matthew Cooper, *The German Air Force, 1933-1945: An Anatomy of Failure* (New York: Jane's Publishing Incorporated, 1981), 65.

[7]Asher Lee, *The German Air Force* (New York: Harper and Brothers Publishers, 1946), 208. Lieutenant General Paul Deichmann stated, "The vast majority of all air missions executed [in 1942] were missions of army support in action directly in front of the German ground forces. Eighty percent of all bomber forces available were employed in missions of direct support for operations on the ground, with only a small number committed against targets in the far enemy rear, in action commensurate with the actual mission assignment. . . . In 1943 80 percent of all air activities were dictated by the mission of tactical support for the Army, and military events in 1944 produced no changes of any consequence in this situation." Paul Deichmann, *German Air Force Operations in Support of the Army*, USAF Historical Studies, No. 163 (USAF Historical Division, Research Studies Institute, Air University, 1962), 164-65.

[8]Lee, 40.

[9]Murray, 112.

[10]Lee, 209.

[11]Williamson A. Murray, "The World in Conflict," in *The Cambridge Illustrated History of Warfare: The Triumph of the West,* ed. Geoffrey Parker (Cambridge: Cambridge University Press, 2000), 303; British Air Ministry, 13; Richard Suchenwirth, *Historical Turning Points in the German Air Force War Effort,* USAF Historical Studies, No. 189 (USAF Historical Division, Aerospace Studies Institute, Air University, 1968), 32.

[12]Lee, 15.

[13]Fritz Morzik, *German Air Force Airlift Operations,* USAF Historical Studies, No. 167 (USAF Historical Division, Research Studies Institute, Air University, 1961), 1; Cooper, 58.

[14]Hugh Trevor-Roper, *Hitler's Table Talk, 1941-44: His Private Conversations* (London: Weidenfeld and Nicolson, 1973), 687.

[15]Cooper, 60.

[16]Ibid., 12.

[17]Captain Jonathan M. House, *Toward Combined Arms Warfare: A Survey of 20th-Century Tactics, Doctrine, and Organization* (Fort Leavenworth, KS: US Army Command and General Staff College, 1984), 55.

[18]Cooper, 9; see also *Strategy for Defeat*, 4-5.

[19]*The United States Strategic Bombing Survey: Summary Report (European War)*, (Washington DC: US Government Printing Office, 1945), 393.

[20]Cooper, 61.

[21]N-9618, *Interrogation of Reich Marshall Hermann Goering, 10 May 1945, 1700 to 1900 hours.* (Combined Arms Research Library, Fort Leavenworth, KS) 4, 10; N-10007-3, *Headquarters Air P/W Interrogation Detachment Military Intelligence Service: Hermann Goering. 1 June 1945* (Combined Arms Research Library, Fort Leavenworth, KS), 21-22; see also, Hermann Plocher, *The German Air Force Versus Russia, 1942,* USAF Historical Studies, No. 154 (USAF Historical Division, Aerospace Studies Institute, Air University, 1966), 2; and Richard Suchenwirth, *Historical Turning Points in the German Air Force War Effort*, No. 189 (USAF Historical Division, Aerospace Studies Institute, Air University, 1968), 73-75.

[22] Suchenwirth, 20, 27; Harold Faber, ed., *Luftwaffe: A History* (New York: The New York Times Book Company, Inc., 1977), 141-142.

[23]Lee, 19- 21.

[24]Ibid., 22-24.

[25]Suchenwirth, 32-35.

[26]Ibid., 34; Faber, 158. Lee points out that from 1943 onwards, the Germans established a unified air-transport command, the 14th Air Corps. Unfortunately for the Luftwaffe, this organization came too late to be of any value; there were too few resources remaining to capitalize on it. Lee, 26.

[27]Lee, 38.

[28]Ibid., 44.

[29]Murray, 312.

[30]*Interrogation of Reich Marshall Goering*, 3.

[31]Suchenwirth, 33; Faber, 157.

[32]Ibid.

[33]Morzik, 7-8.

CHAPTER 2

DANGEROUS PRECEDENTS: DEMYANSK AND KHOLM

It is entirely possible that the Stalingrad airlift may have been avoided if it were

not for the success of two previous German airlifts: Demyansk and Kholm. Demyansk

was the first large-scale airlift operation of World War II,[1] and was a consequence of the

Germans' stalled attack on Moscow, which played into the hands of the Soviet winter

offensive.

In late 1941, the Germans were northwest and south of Moscow attempting to

capture the city when the muddy period began. Heavy frost and snow almost immediately

followed the mud, making movement nearly impossible. Supply operations over land

were very difficult through the frost and snow; cross-country transport of supplies would

become impossible if the cold snap continued. The attack on Moscow ground to a halt.

Offensive operations came to a complete standstill; the only thing left to do was prepare

for defensive operations. Because of the rapid German advance of the vanguard units, the

front line was unbalanced exposing many unit flanks. Severe weather patterns combined

with the difficult terrain and relative immobility of armor and heavy weapons left the

advance units in a precarious situation--withdrawal at the moment seemed undesirable

and impractical since it would have meant a loss of much heavy equipment.[2]

On 9 January 1942, the advancing Soviet winter offensive allowed four Soviet

armies, operating on a sixty-mile front, to penetrate the boundary positions held by two

infantry divisions between German Army Groups North and Center.[3] German troops

were hastily assembled and thrown into towns and villages along the Soviets' path to act

as breakwaters, but the Soviet onslaught continued. Among the attackers were fresh

Siberian troops, operating close to their supply bases and well acquainted with and prepared for the Russian winter.[4] German units, trapped in completely inadequate defensive positions, were overrun and forced to retreat to the west, leaving behind their heavy weapons and vehicles. In the second week of February Soviet forces surrounded all of X Army Corps and parts of II Army Corps, located in the Demyansk area (see figure 1): some six divisions of approximately one hundred thousand men.[5] Within several days the distance between the German front and the enclaves increased to seventy-five miles.[6] Unless the advancing Russian counter offensive could be stopped, the collapse of the middle sector of the front seemed certain.[7]

Demyansk

The only way to overcome the distance, time, and inadequacy of the highway network to supply the isolated forces was by air.[8] Hitler himself ordered the resupply on 8 December.[9] This was a new role for the German Air Force. Rather than paving the way for the army's offensive conquests, they were now rescuing German ground forces in a defensive role.[10] Every aircraft that landed in the operating area with troops, weapons, and supplies would strengthen the German forces, while simultaneously relieving the Soviet pressure.

Colonel Fritz Morzik, chief of air transport, stated that the airlift was possible once certain conditions were met. First, to deliver a daily quota of three hundred tons, he needed at least 150 operational aircraft, since he knew the present strength was 220 aircraft and only one-third of those were serviceable. It was therefore necessary to draw aircraft from other fronts and deplete Germany of all available airlift aircraft. Second, he required additional ground crews and better ground equipment. Especially at Demyansk, the only way to ensure a high degree of operational readiness for an extended airlift was to make certain necessary modifications. Mobile workshops, auxiliary starters, engine-warming carts, etcetera were vital to cope with temperatures of minus forty to fifty degrees centigrade.[11] Third, Colonel Morzik requested elimination of the usual chain of command. This would allow the air transport chief to issue orders directly to ground organization and supply units and to submit requests for needed services and supply items directly to these same agencies. Morzik provided his rationale for the third condition:

> The above requirements were based on the clear realization that an improvised undertaking of the contemplated scope could succeed only if the necessary authority were concentrated in one person. For there is only one agency capable of surveying and integrating the many requirements of a large-scale operation, and that agency is the one charged with the responsibility for directing it.[12]

General Alfred Keller, commander of First Air Fleet, which was responsible for all German Air Force operations in a sector that included Demyansk, agreed to the conditions. Twenty-four hours later, a makeshift airlift was in progress. Ju-52 formations came flying into their new bases at Pleskau-West and South, Korovye-Selo, Tuleblya, Ostrov, Riva and Riga-North, and Daugavpils. The landing airfields were: Demyansk; Pieski; Supply Drop Area, Demyansk; and Kholm. Initially, only two of these airfields had the facilities and equipment to handle large-scale or even routine operations in all-weather conditions: Pskov-South and Riga. The Demyansk airfield consisted of a 2,625-by-164 foot landing strip, a small-taxiing area, and an unloading area that consisted of removing the snow and packing the ground underneath; twenty to thirty aircraft could use Demyansk's improvised facilities at a time.[13] Morzik knew that Demyansk alone would be insufficient to support one hundred thousand men.[14] Inclement weather, aircraft wreckage, or enemy activity could shut down the effort indefinitely. As a contingency, Colonel Morzik demanded a second landing field within the encircled area; Pieske, eight miles north of Demyansk, was completed in March[15] and consisted of a 1,968-by-98-foot landing strip in the hard-packed snow and could support loading and unloading for only three to six aircraft at a time.[16] Morzik limited Pieski operations to the most experienced pilots and restricted loads to one and a half tons to prevent the landing gear from breaking through the snow.[17] Additionally, Supply Drop Area, Demyansk, was a marked, open area used to drop supplies during the muddy period. Kholm could support limited landing activity.[18]

When the operation began, there was no established organization to handle it effectively. Everything had to be flown in: aircraft direction-finding equipment, radio beacons, and even the simplest tools. On 20 February the first forty Ju-52s landed at Demyansk.[19]

A problem that was to haunt the air transport forces throughout the airlift was the lack of integration with the entire Luftwaffe effort. Part of the reason the commander in chief of the First Air Fleet agreed to the air transport chief's conditions was that he did not understand the idiosyncrasies and requirements of a large-scale air transport mission, neither did the air fleet staff. As a result, there was little support provided to the transport forces. In one instance, Colonel Morzik requested information regarding the best approach route into Demyansk to avoid Russian antiaircraft defense. The reply from the intelligence branch of the air fleet staff was to "select that route which offers the best chance of avoiding losses."[20] On another occasion, the air fleet staff failed to inform the air transport staff that the Russians had successfully parachuted forces into the encircled area. The air transport staff learned of this only when a number of transport aircraft returned to home station needing repair after being accidentally hit by German antiaircraft fire.[21] This damage could have been avoided if the air fleet staff had informed the air transport chief of the attack immediately.[22]

Pieski to Demyansk was 150 miles, but the transports were within range of Russian fighters for the final one hundred miles. The Russians would usually lie in wait near Demyansk and attack the unsuspecting transports from abeam as they configured to land. However, if German fighters appeared, the Russians would flee.[23]

Russian efforts to thwart the German airlift hindered Luftwaffe success, despite the unexplained Soviet reticence to attack the German onload facilities. Once the Red Army learned that the airlift missions were part of a systematic and continuing operation, Soviet leaders issued orders to ensure every soldier carried a weapon with him at all times and to fire immediately at any transport aircraft passing overhead. Crews soon reported increased fire from infantry weapons of all kinds. The Luftwaffe resupply effort consequently saw a daily increase in lost aircraft and airmen.[24]

Colonel Morzik initially sent the aircraft out individually at low level, but the Russian flak became too dangerous and increasing numbers of enemy fighters appeared.[25] The steady increase in the loss of aircraft prompted him to increase the enroute altitude to between 6,500 and 8,200 feet.[26] He also opted for transport units to fly in groups of twenty to thirty aircraft to concentrate their firepower if attacked by enemy fighters. Ingress and egress routes were changed each day. These new tactics were initially successful. However, the Soviets adapted their tactics and began to attack the unwieldy transports with single-engine fighters. Usually two to four Russian fighters would attack, but they were not eager to attack large transport groups in close formation, especially when the transports opened fire immediately. Therefore the Russian fighters primarily engaged Ju-52s straggling behind the main group. If the transport pilot was skillful and the gunner was not sparing of tracer fire, the enemy attackers usually fled. Compared to losses by antiaircraft artillery and infantry fire, losses to fighters were few, and some transports even shot down Russian fighters.[27]

Nevertheless, Russian fighters did pose a threat, but because of the vast difference in airspeeds as well as the scarcity of German fighters, there was no plan for a regular fighter escort for transport missions. As a result, Luftwaffe personnel arranged a pre-coordinated time and altitude for the Ju-52s to meet the fighters, which would provide air cover in the encircled area and escort the transports back to friendly airspace. Escort duty was popular, since transports were a lucrative target, enticing Russian fighters to engage. That allowed the Me-109s to attack the Soviet aggressors in defense of the transports, scoring many aerial kills in the process. Although cooperation was excellent, the number of Me-109s in the area varied from two to ten. Luckily, even a token fighter force made the enemy extremely cautious and provided the transport crews with moral support. German fighter cover greatly limited losses to enemy fighters, and was the primary reason why Russian air activity never disrupted the airlift.[28] The danger the Luftwaffe transports faced from Soviet soldiers was no greater than the danger the German soldiers faced from their Soviet counterparts. The German Air Force had to organize the effort to maximize the results for the risks involved.

The size of an airlift required to supply one hundred thousand men, combined with the dimensions at the primary off-load site Demyansk required careful coordination, scheduling, and deconfliction to ensure an orderly flow of aircraft, to prevent overwhelming the off-load areas and personnel, and to avoid having aircraft unnecessarily delayed on the ground, where they would be targets of opportunity for the enemy. Morzik's plan was to give a schedule each day to those units assigned to fly. The schedule designated a specific time for each unit to land at Demyansk, to unload, and to take off for the return flight. The air transport staff computed the timing to allow the units to follow one another in rapid succession, while still avoiding the danger of concentrating too many aircraft over the field at once.[29]

Like most military operations, the Demyansk airlift presented a steep learning curve for all the participants. The inadequacy of ground-support equipment and facilities prompted unit commanders to request immediate action to improve them. The situation could not be fixed immediately, and as a result, the transport forces had to rely on the support of other Luftwaffe units stationed at the same airfields. Original units were reticent. They were having difficulty meeting their own requirements with the difficulties posed by the Russian winter. Colonel Morzik persuaded First Air Fleet Headquarters to reserve at least 50 percent of the facilities to his transport units. The original units complied with great reluctance, while stalling as long as possible since full compliance would jeopardize the success of their own missions. The air transport staff continued to report these deficiencies and urgently requested improvements, for the inadequacy of the facilities threatened the success of the airlift.[30]

Both the air fleet and Air Administrative Command attempted to resolve the situation, but they simply could not accommodate an additional four hundred aircraft within their assigned area. Weeks passed before improvements were made. Operational readiness fell to less than 25 percent of the authorized strength and only one-half of the 150 serviceable aircraft available were able to contribute to the airlift. Morzik explained what must have been the prevailing sentiment among Luftwaffe leaders regarding the decision to execute the air supply:

> Gradually it was becoming obvious to all concerned that the decision to keep an encircled army corps supplied exclusively by air had been based on a completely erroneous, or at least overly optimistic, estimation of the Russian winter, of the resources available to meet technical requirements, and of the insurmountable difficulties inherent in covering the tremendous distances involved.[31]

In addition to the problems with aircraft reliability, the efficacy of the Luftwaffe effort suffered from lack of aviator experience. Many of the pilots involved in the airlift were straight from flying schools.[32] The beginning of the operation was a valuable training period, allowing them to gain valuable experience, while becoming seasoned combat pilots and while learning how to counter the idiosyncrasies of the Russian winter. As the operation progressed and the weather grew warmer, the airfields turned into mud, sharply reducing the adequacy of the taxiing and landing areas at the home bases. Operations were reduced even further in the unloading areas of the landing bases. The unpredictability of field conditions and, consequently, loading and unloading efforts, choked operations. Aircraft took off as soon as they were ready. The varied distances of the departure fields made prior coordination of arrival times nearly impossible. By this time, however, the air transport chief had effective radio communications with Demyansk and was able to relay emergency instructions to airborne aircraft if it appeared that too many aircraft were arriving at once. This system, while far from ideal, was considered a "necessary evil," and facilities at Demyansk were never overwhelmed.[33]

On the return flight over the critical front area, all transport aircraft assembled into groups and remained together until they had crossed the hostile territory, where they split up and returned to their individual bases.[34]

The greatest problem faced by the transport units, however, was the Russian winter.[35] The rubber on the tires tended to get brittle and crack. Fuel and oil pipes would freeze up. Hydraulic pumps broke down. Engines were difficult to start and required constant attention. The water in transformers would freeze, rendering engine instruments and even radios completely unreliable.[36] Remarkably, in spite of the maintenance difficulties as well as days with visibility below 2,000 feet and the ceiling almost to the ground, there was never a single day during the entire operation in which the Luftwaffe was unable to fly airlift missions.[37]

On some levels, the airlift was a success. This was certainly true from a tactical standpoint. Bekker observed that from 20 February, until 18 May 1942, six trapped divisions were kept alive entirely by the air. During this period 24,303 tons of supplies were delivered, a daily average of 276 tons of enough foodstuffs, weapons, and ammunition for one hundred thousand soldiers. Additionally, the encircled army received over five-million gallons of fuel and 15,446 replacements for the 22,093 wounded flown out. Two hundred sixty-five aircraft were lost--less to the enemy than to "General Winter."[38]

Morzik provides a more detailed account of the accomplishments of the air transport forces. From January 1942 until the final clearing of the Demyansk pocket in early 1943, the air transport fleet airlifted 64,844 tons of materiel, weapons, ammunition, gasoline, foodstuffs, spare parts, clothing, medical supplies, mail, and miscellaneous supplies to Demyansk. Thirty thousand five hundred replacement and relief troops were delivered to join their besieged counterparts. Just as important was the evacuation of thirty five thousand four hundred wounded and sick personnel, as well as soldiers departing on leave. From 18 February until 19 May 1942, the airlift provided an average of 302 tons per day, slightly exceeding the 300-ton daily requirement.[39] The air transport forces supplied one hundred thousand men entirely from the air. The beleaguered Germans were relieved by ground forces on 18 May, having been trapped for ninety-one days.[40]

Another factor owing to the success of the airlift was the weakness of the Russian Air Force. Accordingly, air transport losses were low. Throughout the entire airlift, the Russian Air Force never attacked the transport units at their takeoff bases, although parked aircraft, loading and taxing operations, ground-support facilities, supply stores, etcetera, all provided lucrative and enticing targets. Morzik noted the mistake the Soviets made in not targeting the takeoff bases when he observed that:

> A short, tightly concentrated action against the German take-off bases would have enabled the enemy to halt the entire airlift within a very short time, for the transport forces, utilized constantly almost beyond the limit of endurance, had no reserves to fall back on. A well-timed, intensive Russian attack would have effectively sealed the fate of the 100,000 men trapped at Demyansk.[41]

The real cost of Hitler's decision to execute the Demyansk airlift becomes crystal clear when put into a different light. The airlift mission consumed 160 railway trains of gasoline and deprived the pilot training program of three hundred aircraft for a four-month period.[42] However, Morzik said that even these numbers pale in comparison to the dangerous precedent set by the success of the Demyansk airlift:

> The negative aspect of the success at Demyansk was that it led to erroneous evaluation of the status and potential development of a given military

situation insofar as the suitability of air transport was concerned. The potentiality for success of the air transport forces was viewed with far too much optimism and, from this point of view, the Demyansk operation must be considered a turning point for air transport.[43]

As previously mentioned, the Demyansk airlift was a tactical success, but what were its strategic implications? Achieved only at tremendous cost, its superficial success made reliance on air supply a "stopgap solution."[44]

Kholm

The airlift operation at Kholm was a subordinate part of the overall Demyansk operation.[45] The Red Army had surrounded three thousand five hundred men in a pocket only 1.25 miles in diameter.[46] This area was too small for an airstrip, so the Ju-52s had to land on a snow-covered field in no-man's-land, to drop the supplies while taxiing, and then to take off before the Russian artillery could open fire. High casualties halted this undertaking in favor of resupply from airdrops by He-111s and heavy gliders, which landed with the supplies in front of German lines.[47]

Enemy activities at Kholm differed from those at Demyansk. At Kholm, Soviet forces gradually moved in closer and closer, making the air supply more and more difficult.[48] Despite fervent efforts by the Germans, the Russians slowly gained ground. Initially, the Ju-52s and gliders could land at the airfield despite enemy artillery fire. Eventually, Russian intermediate and light infantry fire made it impossible to land there, mandating the use of airdrops as the only practical method of manned resupply.[49]

He-111s would airdrop their cargo and use their weapons to protect the heavy gliders that were in bound to Kholm, where they had to land in the middle of a city street.[50] The street was in front of German lines; the German soldiers would rush out to collect the vital cargo. Occasionally the Russians got there first, but enough supplies reached the encircled Germans that they were able to continue resisting until they were liberated in early May.[51]

---

[1]Fritz Morzik, *German Air Force Airlift Operations* (Air University: USAF Historical Division, June 1961), 137.

[2]For a detailed account of the Soviet winter offensive, see John Erickson, *The Road to Stalingrad: Stalin's War with Germany, Vol. 1* (London: Harper and Row, Publishers, 1975), 249-96. For detailed accounts of the ground situation leading up to and during the events at Demyansk and Kholm, see Erickson, 279-381.

[3]Cajus Bekker, *The Luftwaffe War Diaries: The German Air Force in World War II*, trans. and ed. Frank Ziegler (New York: Da Capo Press, 1994), 275.

[4]Morzik, 139-A.

[5]Bekker, 275; Plocher, 78.

[6]Ibid.

[7]Morzik, 139-A.

[8]Ibid.

[9]Richard Muller, "The German Air Force and the Campaign Against the Soviet Union, 1941-1945" (Ph.D. diss., The Ohio State University, 1990), 108.

[10]Asher Lee, *The German Air Force* (New York: Harper and Brothers Publisher, 1946), 119.

[11]Bekker, 276.

[12]Morzik, 151.

[13]Ibid., 149.

[14]Bekker, 276.

[15]Ibid.

[16]Morzik, 149.

[17]Bekker, 276.

[18]Morzik, 149.

[19]Bekker, 276

[20]Morzik, 151-152.

[21]Hermann Plocher, *The German Air Force Versus Russia, 1943,* USAF Historical Studies, No. 155 (USAF Historical Division, Aerospace Studies Institute, Air University, 1967), 82.

[22]Morzik, 152.

[23]Bekker, 276

[24]Morzik, 157-158.

[25]Bekker, 276.

[26]Bekker states that the altitude was 6,000 feet. Bekker, 276.

[27]Morzik, 158.

[28]Ibid., 159-160.

[29]Ibid., 153.

[30]Ibid., 161

[31]Ibid., 162.

[32]Bekker, 277.

[33]Morzik, 155.

[34]Ibid., 155.

[35]Bekker, 276.

[36]Ibid., 277; Morzik, 164-165.

[37]Ibid., 169.

[38]Bekker, 277.

[39]Morzik, 172. The reason behind the discrepancy between Bekker's and Morzik's figures is unknown. Many figures in various Luftwaffe operations conflict, depending on the source. There are several possibilities as to why. The sending team recorded what was sent, the receiving team recorded what was received. Their biases or differences in scales or other equipment could produce variations. If an aircraft was lost,

diverted, shot down, crashed, etc., the numbers would not add up. Regardless of the difference, the result is that the operation was a *tactical* success.

[40]Matthew Cooper, *The German Air Force, 1933-1945: An Anatomy of Failure* (New York: Jane's Publishing Incorporated, 1981), 241; the airlift consisted of 32,427 supply missions and of 659 missions limited to personnel. The entire operation consumed 42,155 tons of aviation gasoline and 3,242 tons of lubricants. In addition to the aircraft losses stated above, 383 officers, noncommissioned officers, and enlisted personnel were either killed, wounded, or missing in action. Morzik, 172.

[41]Morzik, 160.

[42]Ibid., 143

[43]Ibid.

[44]Ibid., 179.

[45]Ibid., 173-A.

[46]John Erickson said there were 5,000 men trapped at Kholm. Erickson, 306.

[47]Cooper, 241; Bekker, 277; Plocher, 74.

[48]Morzik, 173-A.

[49]Ibid.

[50]Ibid., 174.

[51]Bekker, 277.

CHAPTER 3

STALINGRAD

> Where the German soldier once stands there he remains
> and no power on earth will drive him back.[1]

Adolf Hitler, *Stalingrad*

The success of the Demyansk and Kholm airlift set dangerous precedents.[2]

German leadership now possessed a confidence in their Luftwaffe that may have been

partially responsible for the entombment of General Friedrich Paulus' army at the Battle

of Stalingrad. Events that precipitated the Stalingrad airlift were twofold: the Russian

winter and a Soviet counterattack. In late 1942, General Paulus's Sixth Army, composed

of about 250,000 men, was fighting to capture the industrial city of Stalingrad.[3] Seven-

eighths of the city was already captured when the onset of winter coincided with a Soviet

counterattack on 19 November.[4] Two days later the Sixth Army found themselves on the

horns of a dilemma. They could make a fighting retreat or allow themselves to become

separated from the main German front and trapped between the Don and the Volga.

Paulus was not inclined to retreat, concerned that he did not possess enough fuel to

succeed. However, Paulus's misgivings became irrelevant once Hitler decreed that the

Sixth Army was to defend Stalingrad under any circumstances.

With the Sixth Army aware of the pincers movement forming around them by

Soviet armor, Lieutenant General Martin Fiebig, commander of VIII Air Corps during the

Stalingrad operations on 21 November, telephoned the Sixth's Army' chief of staff,

Major General Arthur Schmidt. With Paulus himself listening to the conversation, Fiebig

asked Schmidt what the army's plans were:

"The C.-in-C.", answered Schmidt, "proposes to defend himself at Stalingrad."

"And how do you intend to keep the Army supplied?"

"That will have to be done from the air."

The *Luftwaffe* general was flabbergasted. "A whole Army? But it's quite impossible! Just now our transport planes are heavily committed in North Africa. I advise you not to be so optimistic!"

Fiebig promptly reported the news to his *Luftflotte* chief, Colonel-General von Richthofen, whose telephone call in turn woke up the chief of general staff, Jeschonnek, at Goldap.

"You've got to stop it!" Richthofen shouted. "In the filthy weather we have here there's not a hope of supplying an Army of 250,000 men from the air. It's stark staring madness!"[5]

But the memories of the airlift successes at Demyansk and Kholm still resonated strongly and fate took its course.

Historians have argued the soundness of the decision to execute the airlift, as well as who really is to blame for making that decision. Since there was no German air transport fleet, there were no high-ranking airlift experts to advise the decision makers on the soundness of any decision regarding large-scale airlifts, nor to present cost-benefit analyses from previous airlifts to predict the risks associated with subsequent efforts under different circumstances. Weather conditions, enemy strength and activity, preexisting airlift infrastructure, and other factors can all pose significant challenges and increase the risks for different operations. For the Stalingrad airlift, the status quo since Demyansk and Kholm had changed.

In hindsight, the order to execute the Stalingrad airlift was disastrous. Hitler was the ultimate authority and the primary responsibility rests with him.[6] However, Joel Hayward concluded that there were three people to blame: Hitler, Göring, and Jeschonnek. Jeschonnek, initially believing that the supply would be a temporary operation to allow Paulus to break out, rashly promised Hitler that the Luftwaffe was capable of meeting the army's needs, before he had consulted with airlift experts, made his own calculations, or spoken with General Wolfram von Richthofen (commander of the Fourth Air Fleet) and the other air force and army commanders at the front.[7] Had he taken any of these actions, it is possible that he would have cautioned Hitler as to the possibility of failure of an airlift. Later, when he learned the airlift was to be considerably longer and Richthofen and Fiebig were strongly opposed to the airlift, he admitted his mistake and attempted to dissuade Hitler and Göring.[8]

Hayward cited seven reasons that Hitler's blame is greater than Jeschonnek's. Only four will be discussed here. First, his egotism caused him to believe that his iron will is what saved the eastern armies the previous winter and would do so again. He did not look objectively at the situation facing the Sixth Army. Second, he would lose face if he allowed a withdrawal after publicly promising to keep the city. Third, he turned a deaf ear to the repeated pleas and warnings of his frontline army and air force commanders, calling them defeatists for questioning his stand fast solution that he had elevated to doctrine. Fourth, he did not fire Göring and replace him with someone competent despite the Reichsmarschall's poor track record and the negligence of his command the previous year or even insist that Göring act responsibly during this crucial period, rather than in the haphazard fashion for which he was known.[9]

Göring's responsibility for the airlift decision was equal to Hitler's. When the Nazi leader asked him if the Luftwaffe could meet the Sixth Army's needs as Jeschonnek had promised, he should not have blindly agreed. He should have first consulted his airlift experts; sought the opinions of Richthofen and his other commanders; and thoroughly acquainted himself with the conditions at Stalingrad to include the enemy order of battle, size and needs of trapped forces, weather patterns and conditions, and operational readiness of Fourth Air Fleet. Göring did none of these things. His assurances to Hitler may have been nothing more than an attempt to restore his importance and influence with the führer.[10]

Luftwaffe commanders in the field were united in their belief that the air force could not supply the entire Sixth Army and in their objection of the idea to local army commanders and the High Command itself. Fiebig's thoughts on the airlift's feasibility have already been mentioned. Richthofen urged an immediate breakout. He noted in his diary on 21 November, "Sixth Army believes that it will be supplied by the air fleet in its hedgehog positions, I made every effort to convince it that this cannot be accomplished, because the necessary transport resources are not available."[11]

The next day Major General Wolfgang Pickert, the 9th Flak Division commander, repeated theses same thought to Paulus and Schmidt during a conference. Pickert insisted a breakout was the only option. When Schmidt asked him what he would do, he replied, "I would gather together all the forces I could and break out to the southwest."[12] Schmidt replied that the Sixth Army would not attempt a breakout. Hitler had expressly forbidden a breakout and the enemy held the high ground to the west, thus exposing the Sixth Army to Soviet guns if they did attempt a breakout.[13] The Sixth Army would remain in the pocket and defend themselves as Hitler had ordered. Now decisions had to be made for the organization and leadership of the airlift.

German leaders placed responsibility for the airlift with the Fourth Air Fleet, commanded by von Richthofen. Major General Wolfgang Pickert was the senior Luftwaffe officer inside the pocket and was responsible for the effort to receive the supplies and defend the airspace around Stalingrad. Initially, Fourth Air Fleet appointed Major General Victor Carganico, Commander of the Airfield Area Tatsinskaya, as the Stalingrad Air Supply Chief. It soon became apparent that General Carganico and his staff were in over their heads having insufficient airlift experience.[14] On 29 November, Fourth Air Fleet relieved Lieutenant General Martin Fiebig, commander of the VIII Air Corps, of his combat-mission responsibilities and directed him to assume responsibility as the Stalingrad Air Supply Chief[15]. He possessed an experienced command staff as well as communication and weather facilities, and fighter and bomber units for escorts. Colonel Foerster took command of the air transport units then assigned to the Air Fleet at Tatsinskaya.

Fiebig's assignment resulted in the use of bombers for airlift, but those same aircraft were also desperately needed for bombing missions to support the fighting along the front. At this time, Colonel Ernst Kuehl received command of the He-111 units, stationed at Morozovsk, and became Air Transport Chief (Morozovsk). Colonel Foerster retained command of the Ju-52 units at Tatsinskaya, but was later replaced by Colonel Morzik. Major Willers became Air Transport Chief (Stalino) and assumed command of the long-range aircraft stationed there.

The Quartermaster General of the Sixth Army at Morozovsk submitted his requests for provisions to Army Group Don. The Army Group then arranged for the transportation and rigging of the required items. An Army Group liaison officer at each airfield ran a supply detail charged with packing and loading the aircraft, and also assisted the medical staff with evacuation. Eventually, coordination among these agencies was good.[16]

Sources vary, but estimates to sustain a fighting force of 250,000 men range between six hundred and 750 tons per day. The Sixth Army's supply requirements were initially established at 750 tons per day, but later reduced to five hundred tons per day.[17] The required aircraft and crews for the Stalingrad airlift assembled on short notice from the advanced flight training school.[18] Sending many of the Luftwaffe's most experienced instructor-pilots contributed to degradation in the quality of new pilots being trained. If the best instructors were removed from flight training and deployed to the front, who was left to train future aviators to allow sustained operations and maintain combat effectiveness?

Every single available aircraft mobilized for the Stalingrad airlift. On 23 November, Lieutenant General Hans-Georg von Seidel, the Quartermaster General of the Luftwaffe, ordered all Ju-52s (transport aircraft); Ju-86s (trainer; completely inappropriate as a transport); FW-200s and Ju-90s (long-range reconnaissance aircraft); He-111s (long-range bomber), from every unit, staff, ministry, and the Office of the Chief of Training. Six hundred aircraft along with some of the best flight instructors were stripped away from the training facilities. Specialized training schools were closed due to the ruthless efforts taken to ensure the success of the airlift.[19] By early December, Fourth Air Fleet had approximately five hundred aircraft at their disposal, with more becoming available as the operations progressed. Germany's top military leaders were convinced that the number of aircraft now dedicated to the operations was sufficient to meet the logistical needs of the Sixth Army.[20]

Military leaders were incorrect if they assumed that the requisition order of 23 November would mean the miraculous arrival of all available Ju-52s a mere twenty-four hours later. Korovograd and Zaporozhe airfields were supposed to winterize the aircraft. However, personnel assigned to these bases were more concerned with reporting turn-around statistics than with properly preparing the aircraft for the Russian snow and ice. Accordingly, the aircraft arrived at Tatsinskaya ill prepared for winter employment and choked the airfield awaiting proper equipment.[21] Lack of winter preparation for the transports was only one problem the Luftwaffe faced in an effort to maximize throughput to Stalingrad; the human dimension was also overlooked.

Aircrews did not have a certification program, local area familiarization, or any other method of adapting to the exigencies present at Stalingrad. They were fully employed upon their arrival. Morzik states that the bitter cold; perilous approach and return flights facing heavy enemy fighters and antiaircraft artillery; steady shelling of the home and offload bases; ground operations within the enclave while facing constant artillery fire and grenades; and the ubiquitous danger of icing and the mechanical hazards caused by the severe cold presented a difficult challenge. The crews also had to confront the psychological depression inherent when dealing with starving troops and countless wounded, many of whom were left behind owing to lack of space to fly them out. These young and inexperienced crews were deeply disturbed by their experiences.[22]

The only two acceptable airlift bases, Tatsinskaya and Morozovskaya, were located 160 and 130 miles (60 and 50 minutes' flying time)[23] respectively from the encircled airfield of Pitomnik, a distance that did not allow the aircraft to trade much fuel for freight. The Ju-52 could carry about two and a half tons, and the He-111 could carry only two tons. Spread between Tatsinskaya and Morozovskaya, a total of approximately 350 Ju-52 and Ju-86 transports and 190 He-111 bombers were available for the airlift.[24]

37

Despite this ostensibly adequate airlift fleet, other factors proved to be far more significant in determining the Sixth Army's fate than simply the sheer number of aircraft. An army requirement of 600 to 750 tons per day would require 240 to 300 missions per day. These numbers are based on the optimistic assumptions that the airlift could and would operate nonstop, day and night, with no allowance for airfield capabilities, mechanical difficulties, or other conditions that would affect operational readiness rates and delivery amounts; that aircraft would be able to carry their maximum designed cargo weight, with no consideration for fuel-versus-cargo limitations; that the enemy would not interfere; and that the barbaric Russian winter would not hinder the operation. The 240 to 300 missions per day numbers are based on the two-and-a-half ton cargo capacity of the Ju-52.[25] Considering a majority of the airlift aircraft were less capable than the Ju-52, the mission was doomed before it started. Even if adequate delivery platforms were available, the Germans could not begin to keep up with the offloading operations, because 240 missions per day would equate to one aircraft every 6 minutes, assuming the Germans could maintain twenty-four hour operations; three hundred missions equals one landing every 4.8 minutes! Under ideal conditions, this is possible. However, taking into account the Russian weather and Luftwaffe countertactics to mitigate the Soviet threats, many times the transports were forced to arrive in large numbers, hopelessly overwhelming offloading ground crews and their equipment. The "train wreck" on the ground was probably equal to the chaos created once additional inbound aircraft discovered they were unable to land due to a clogged runway and airfield, and were forced to return unexpectedly to their home base. Thus, without the ability to maintain a consistent, sequential, and predictable (at least to the Germans) airlift flow, the operation was doomed before it even started, regardless of the number of aircraft dedicated to the airlift.

There were six airfields available in the pocket; but Pitomnik and Basargino were the only significant airfields (see figure 2).[26] Only Pitomnik was capable of handling large-scale operations, in addition to being the only airfield in the pocket with night capability. The others--except Basargino, which Pickert equipped with minimum requirements--were nothing more than bare-grass landing strips, lacking the necessary radio and air traffic control equipment. Several of those fields had been used previously to supply the Sixth Army, but the weather had been better, and the loads smaller.[27]

According to Vaughan and Donoho in their detailed study of the use of tactical airlift to support isolated land battle areas, the Germans at Stalingrad would have needed twenty-five airfields to effectively resupply the besieged troops. Their conclusion also suggests that a protracted combat operation demands at least one runway per ten thousand combat personnel. Granted, at Stalingrad, there were two runways, but Vaughan and Donoho point out that Gumrak was too small to be of any practical use.[28]

In addition to the inadequate number of airfields in the pocket, the Germans failed to account for the atrocious and bitterly cold weather conditions, as well as the uncertainties and hazards associated with operating airlift aircraft in a war zone. Unbeknownst to the Germans, the Volga is "meteorological frontier," marking the boundary where cold, Siberian air from the steppes of Asia collides with warm, moist air from the Black and Caspian Seas.[29] In addition to the seemingly endless days of snow, clouds, and fog generated from this clash of weather patterns, the Luftwaffe had to endure brutal temperatures as low as fifty degrees Fahrenheit below zero.[30] Brigadier

General Wolfgang Pickert, the flak commander and senior Luftwaffe officer trapped in

the pocket, later recalled:

> The cold caused unimaginable difficulties in starting aircraft engines, as well as engine maintenance, in spite of the well-known and already proven "cold starting" procedures. Without any protection against the cold and the snowstorms, ground support personnel worked unceasingly to the point where their hands became frozen. Fog, icing and snowstorms caused increasing difficulties, which were compounded at night.[31]

In addition to the bitter cold, Luftwaffe forces were forced to stand down for days

on end due to ice fog, heavy snow, and other weather factors making flight impossible.

There was no respite for the crews and ground personnel. Many times delays were

unavoidable. If the Ju-52s were packed for an airland mission, and the unpredictable

weather changed to preclude landing in the encircled area, then standby ground crews

would have to rig the supplies for airdrop. They had to be ready to go at a moment's

notice to take advantage of any favorable weather.

When the weather did allow airlift operations, there was always the threat of

Russian flak and fighters. Initially, Russian fighters were surprisingly acquiescent toward

the German transports, but as the weeks passed, fighters became more and more active.

Russian fighters forced the transports to fly in groups of forty or fifty in order to

maximize the efficiency of German fighter escort.[32] Simultaneously, ground operations

required the use of German fighters and bombers to repel heavy Soviet ground attacks.

All these factors continued to add to the attrition toll.

At the beginning of the airlift, with good weather and high ceilings, supply units

flew in squadrons or in groups of five aircraft with a fighter escort. During times of low

visibility or low cloud ceilings, only crews fully proficient with instrument flying flew in

groups of five, the rest flew in groups of two of three. Night missions were always flown individually, necessitating carefully coordinated takeoff schedules between Morozovskaya and Tatsinskaya.[33]

Throughout the entire operation the Russians attacked the transports from the air and the ground. Pitomnik possessed relatively strong antiaircraft artillery, yet these weapons did not deter the Russian attacks. Personnel and aircraft losses were high, particularly when the attacks came during takeoff and landing operations, or when the transports were being loaded or unloaded. The relentless bombs, artillery fire, and even grenade attacks took considerable tolls on the operation and even brought it to a standstill at times.[34]

Tatsinskaya and Morozovskaya were the largest and best-equipped airfields in the region, as well as the principal supply bases for Stalingrad; their loss would be a tremendous setback to the airlift and, accordingly, Soviet bombers attacked them repeatedly.[35] On 21 December, two armies of Russian Guards had broken through Axis defenses and were heading south toward Rostov. The Twenty-fourth Tank Corps had advanced to within twenty kilometers of Tatsinskaya, while Soviet bombers were pounding Morozovskaya. The whole German southern front was in danger, but the immediate objectives were Tatsinksaya and Morozovskaya.[36] General Fiebig had requested permission to evacuate before the Red Army was in a position to fire upon it. At first, Richthofen told him to stand fast. He would seek clarification from the High Command. A day later, High Command had still given him no reply. Acting on his own accord, he ordered Fiebig to prepare both Tatsinskaya and Morozovskaya for immediate evacuation should enemy forces threaten the airfields. Early evacuation was essential to ensure the availability of important equipment for future missions. He could not afford to lose fuel tankers, engine-warming equipment, and spare parts, which were in dangerously short supply.[37]

On 23 December, Göring himself finally stepped-in and flatly refused to allow the 180 Ju-52s to retreat until they were under direct fire.[38] An entire transport fleet, representing the merest hope of survival for the Sixth Army, was at stake.[39] Fiebig accepted the orders with the same reticent acquiescence shown by Paulus at Stalingrad. That evening Fiebig wrote in his diary: "I see that we're rushing headlong into disaster, but orders are orders!"[40] As a result, Fiebig refused to allow the evacuation even after Soviet artillery batteries and tanks began shelling the field. A tank shell destroyed the signals center.[41] Artillery and tank gunfire destroyed several transports on the ground, but this did not prevent other aircraft from taking off, despite a visibility of less than 2,000 feet and a ceiling of less than 100 feet.[42] Bekker describes the event that finally made Fiebig relent and order the evacuation on his own authority. Colonel Herhudt von Rohden, Fourth Air Fleet's Chief of Staff, stood beside Fiebig in his shelter, up to this time remaining silent as to what was unfolding before him. An hour and a half after the shelling began, Fiebig's chief of staff, Lieutenant Colonel Lothar von Heinemann, burst into Fiebig's shelter after witnessing the pandemonium breaking-out amongst the crews:

> "*Herr General*," he panted, "you must take action! You must give permission to take off!"
> "For that I need *Luftflotte* authority canceling existing orders," Fiebig countered. "In any case it's impossible to take off in this fog!"
> Drawing himself up, Heinemann stated flatly: "Either you take that risk or every unit on the airfield will be wiped out. All the transport units for Stalingrad, *Herr General*. The last hope of the surrounded 6th Army!"
> Colonel von Rohden then spoke. "I'm of the same opinion," he said.
> Fiebig yielded. "Right!" he said turning to the *Gruppen* commanders. "Permission to take off. Try to withdraw in the direction on Novocherkassk."[43]

Aircraft took off in all directions. Two collided directly over the field. Other taxied into each other, bumped wings on takeoff, or damaged their tail assemblies. The aircraft, those that were still flyable, scattered. On 28 December, the Germans recaptured the airfield, but only for several days as their time was running out.

Fiebig's disobedience saved the transport fleet from certain annihilation. One hundred eight Ju-52s and sixteen of his Ju-86s made it to various airfields, but he lost nearly one-third of his operational aircraft.[44] Aircraft operational readiness rates sank to less than 25 percent.[45] Perhaps just as significant was the loss of vital ground equipment to ensure the surviving aircraft would remain operational at their new base, because Göring's decision to forbid a preemptive evacuation meant that when the moment to escape arrived, the aircraft were still heavily loaded with boxes of ammunition and canisters of fuel for the trapped forces at Stalingrad.[46]

Twenty-miles further east, Morozovskaya was under the same threat. Colonel Ernst Kuehl realized the danger and did not hesitate. When he received the first telephone call that Tatsinskaya had been overrun, he ordered his He-111s and *Stukas* to depart for Novocherkassk. He remained behind and hoped for flying weather that would allow his bombers to keep the Red Army in check.[47] Christmas Day the weather cleared; Colonel Kuehl's forces returned to Morozovskaya and turned back the Soviet spearheads. The airfield was safe for the time being, but the fair weather gave way to ice and fog allowing the Soviet armor to resume their attack. By early January 1943, both Tatsinskaya and Morozovskaya were finally abandoned.[48]

The loss of Tatsinskaya and Morozovskaya forced the Luftwaffe to operate from bases sixty miles farther from the pocket, which retarded the delivery rate and used up precious fuel intended for the Sixth Army. The Ju-52s resettled at airfields in Salsk, located in the northern Caucasus, 250 miles from Stalingrad.[49] This distance was very close to the maximum range of the transports and aircraft that consumed aberrant amounts of oil were sent home.[50] The He-111s now operated from Novocherkassk and were hampered by the same distance problems as the Ju-52s. The new distance was 205 miles, an increase of eighty miles.[51] The longer routings correspondingly cut even deeper into the supplies available to those depending upon them for their survival. But the Russians also had some tricks up their sleeves to further complicate transport efforts.

Capitalizing on the increased flight distance, the Russians set up a continuous line of flak sites along the Pitomnik radio beam, forcing the transports to take longer routings to avoid the deadly ground fire. These necessary detours exacerbated an already critical fuel shortage.[52] By the end of December, only 375 German single-engine fighters existed along the entire Eastern Front,[53] forcing them to be thinly spread and unable to bring any concentration to bear. Meanwhile, Russian fighters grew stronger and stronger as the operation progressed. As the German line was forced farther and farther back from Stalingrad, the city fell beyond the effective range of the fighters.[54] The paucity of German fighters available for escort forced the transports to fly in groups of forty or fifty to counteract the resurgent Red Air Force.[55] Regardless of these measures, the longer distances meant the fighters were unable to escort the airlifters the entire distance, leaving them vulnerable to enemy fighter attack as they approached the pocket.[56] Another tactic to minimize detection and destruction from enemy fighters was to approach the airfields with large waves of transports arriving simultaneously from different directions. This method usually started with a flight of two or three Ju-52s under fighter escort (at least while fighters were still available) approaching the field from a different direction than the main body of transports.[57]

Nevertheless, the Soviets were a thinking, adapting enemy. They witnessed first-hand the operations at Demyansk and Kholm the previous winter, and had learned the requisite lessons to greatly improve their airborne barricade.[58]

Meanwhile, the hemorrhage of aircraft and personnel continued. The rate of attrition always exceeded the rate of replacement, regardless of the additional transport aircraft and personnel robbed from the schools and training installations.[59]

In an effort to prevent unnecessary bloodshed, on 8 January the Soviet High Command issued an ultimatum to the Sixth Army to surrender.[60] Paulus, still maintaining blind obedience to Hitler, refused.

The Russians continued to attack with ever-stronger forces. It seemed only a matter of time before Salsk would be overrun. As a precaution, the Germans built a 2,000 by 100-foot landing strip in the snow at Zverevo, a cornfield lying along a railroad line north of Shakhty, just barely within the operational range of the Ju-52s.[61] On 18 January, Soviet bombardment and strafing attacks damaged thirty Ju-52s, ten completely and the other twenty required repairs out of the theater. This attack precipitated the move from Salsk to Zverevo.[62] Under the direction of Colonel Morzik, the leader of the Demyansk airlift of the previous winter, the Ju-52s began operations from Zverevo. However, within 24-hours, Russian bombers attacked Zverevo as well. Morzik lost fifty-two aircraft, twelve completely and forty damaged.[63]

The most efficient way to run a sustained airlift operation is to have regularly scheduled missions, with enough time between landings so that organic ground personnel and offloading equipment can *immediately* begin and *complete* the offload prior to the arrival of the next aircraft. Of course, if there are sufficient ground crews and equipment to support simultaneous offloading operations, the schedule can be adjusted to reflect this capability. However, in a hostile environment, planners need to stagger arrival times to maintain an element of unpredictability to decrease the enemy's ability to anticipate arriving aircraft. In either case, planners need to meticulously deconflict inbound aircraft. Otherwise, the offload site becomes saturated, forcing additional arrivals to wait short of the offload site, increasing aircraft ground time, which, in turn, increases exposure to enemy threats. If the aircraft arrive too rapidly, it is possible that there will not be room for additional aircraft to land. If the inbound supplies are not critical, those aircraft can return home, wasting only a sortie (or sorties, depending on the number of aircraft) and jeopardizing an aircraft and crew over enemy territory for no reason. If, however, the supplies must be delivered, the aircraft are compelled to orbit, where they will be at risk to enemy aircraft, artillery, and infantry, until there is sufficient room to land. It was this exact scenario that played out at Pitomnik.

With the Red Air Force becoming more active as each day passed, the transports were obligated to fly in groups of forty or fifty. Ground crews at Pitomnik would sit around for hours with nothing to do, then suddenly they would be overwhelmed with forty or fifty aircraft, all desperate to be unloaded immediately, which was not possible, and precious time was wasted.[64]

To make matters worse, the Soviets overran Pitomnik airfield 16 January. Now the Russians had free access to the Pitomnik landing apparatus: German airfield lighting and direction-finding equipment. They set-up a decoy installation and beguiled several German pilots into landing in the middle of Soviet forces.[65]

The fighters assigned to Pitomnik for base defense had performed remarkably, not allowing a Russian fighter or ground attack aircraft to stop the arrival of supplies or the evacuation of forty-two thousand wounded.[66] With the Russians at the doorstep and destruction imminent, the six *Stukas* and six Me-109s[67] stationed there were ordered to evacuate to Gumrak. Unbeknownst to the German pilots, the Sixth Army had not prepared the airfield at Gumrak, despite the requests of the VIII Air Corps since the beginning of the airlift.[68] The airfield was in such poor condition with bomb craters and snow drifts that five of the six Me 109s were destroyed in their attempts to land.[69] From this point forward, the Luftwaffe lost air superiority over Stalingrad and its approaches.[70]

Gumrak had been a Russian Army airstrip. Pickert said that it was "inoperable at the beginning of the siege due to aircraft wreckage, trenches, and artillery craters from earlier battles, as well as deep snow."[71] Richthofen actually wanted the Gumrak airfield improved and enlarged several weeks earlier and he ordered Luftwaffe personnel within the pocket to prepare it for the airlift operations, but the Sixth Army refused permission. Their headquarters were there, along with hospitals, supply installations, and several command posts, so they did not want any construction that might attract Soviet attention to their location.[72] With Pitomnik out of the picture, Gumrak was the only airfield left within the pocket. Among other problems owing to Paulus' refusal to prepare the airfield, the pilots had difficulty finding the runway, because it had no radio beacon.[73] Fiebig noted in his diary on 16 January, "Now we're paying the price for that decision."[74]

By mid-January, the aircraft reliability rate was critical. On 18 January, less than 7 percent of Ju-52s were available, 33 percent of the He-111s, 0 percent of the Fw-200s, and 35 percent of the He-177s.[75]

The situation grew more and more desperate. Before Pitomnik fell, the Sixth Army had already appealed to Berlin for help, which only increased Hitler's nervousness. In an attempt to salvage *something*, Hitler sent Field Marshall Erhard Milch to Fourth Air Fleet to take over the airlift and gave him special powers and authority to issue orders and take any action he deemed necessary for the armed forces in the region.[76] But what could Milch do that other Luftwaffe officials could not?

Milch discovered upon arriving at Richthofen's headquarters at Taganrog airstrip that he possessed fifteen operable Ju-52s out of 140 total, and forty-one operational He-111s out of 140, and one operational FW-200 out of twenty. With VIII Corps' three hundred total aircraft, he had fifty-seven that could fly, or 19 percent.[77] At the time of his arrival, Ju-52s were unable to land within the cauldron. They had resorted instead to airdropping canisters of supplies, and even to flying past the troops and merely pushing the supplies out of the open doors.[78] He raised operational ready rates and supplies for the suffering troops to 30 percent, which was still too low to save Paulus' army,[79] but the cost was additional losses of Ju-52s and He-111s, as well as the irreplaceable instructor pilots who were shot down with their aircraft.[80] Milch even planned to use gliders, but slowly changed his mind after realizing that Richthofen and Fiebig were right. Gliders were not suitable for the conditions at Stalingrad.[81]

Despite reports by Sixth Army to the contrary, the Luftwaffe considered Gumrak unsuitable for transport aircraft; the snow was too deep and Sixth Army troops, weakened

48

from starvation, were too weak to pack it down enough to support landing impact.[82]

Pickert described the degradation in the airlift organization as a result of losing Pitomnik and Bassargino, where he considered the transfer of supplies to be efficient, when he stated that their loss:

> caused some early off-loading difficulties at Gumrak; however a makeshift organization quickly came into being for the few days during which the field was in use. Furthermore, one must not overlook the fact that no equipment was available for snow removal, ground leveling and the removal of aircraft wreckage and other debris. Everything had to be done in a makeshift manner with a few trucks and with manual labor, that is, shovels in the hands of exhausted men.[83]

Even though the Luftwaffe assessed conditions at Gumrak as unsuitable, the Sixth Army radio messages claimed that the airfield was "day-and-night operational."[84] Many transport pilots were convinced the airfield was too dangerous to land and had resorted to merely throwing out supply canisters, if they flew over the field at all.

Aircrews also had to unload their own aircraft. Determined to discover the true status of the field, on 19 January VIII Air Corps[85] sent their representative, Major Erich Thiel, commander of an He-111 bomber group that had been converted to an improvised transport role, into Gumrak. He landed in an He-111 to assess the condition of the runway and offload operations and then report his findings back to his superiors. Milch wanted him to contact Paulus in an effort to convince the latter to improve the conditions at the airfield.[86] The army leader refused to accept any criticism for the ground operations, even when Thiel reminded him that aircraft turnaround time had slipped to around five hours; even when Thiel pointed-out that the airfield, including the runway, was littered with wrecks, Paulus still claimed that it was not his responsibility.[87]

Thiel's report concluded that aircraft were cleared for landing during the day despite the thirteen aircraft wrecks littering the field but only the most experienced crews

49

could land at night. Of particular concern for heavily laden transports attempting a night landing was the wreckage of an Me-109 at the end of the runway. The field was exposed to enemy fighters, which circled the field at 2,500 to 3,000 feet when the weather was clear. Ju-52 landings would be impossible when enemy fighters were present unless the weather was bad. Enemy artillery also threatened safe operations. The airfield was also strewn with unrecovered airdrop canisters half-buried in snow. Regarding the offloading procedures, Thiel added that he landed at 1100; by 2000, he had not even seen an offload team. By 2200, his aircraft still had not been unloaded or defueled despite the fact his aircraft was carrying excess petrol for the fuel-starved army. Airfield personnel claimed the reason was the constant shelling. Other aircraft were unloaded by their own crews, where the supplies were left unguarded and then stolen by passing soldiers.[88]

Major Thiel reported his conclusions to Paulus who, in the presence of several staff officers, then replied:

> When [aircraft] don't land, it means the army's death. It is too late now, anyway. . . . Every machine that lands saves the lives of 1,000 men. . . . Dropping [supplies] is no use to us. Many supply canisters ["bombs" in Paulus's original text] are not found, because we have no fuel with which to retrieve them. . . . Today is the fourth day in which my troops have had nothing to eat. We could not recover our heavy weapons [during recent withdrawals], because no fuel was available. They are now lost. The last horses have been eaten. Can you imagine it: soldiers diving on an old horse cadaver, breaking open its head and devouring its brain raw? . . . What should I say, as supreme commander of an army, when a man comes to me, begging: "*Herr Generaloberst*, a crust of bread?" Why did the Luftwaffe say that it could carry out the supply mission? Who is the man responsible for mentioning the possibility? If someone had told me that it was not possible, I would not have reproached the Luftwaffe. I would have broken out.[89]

Responding to criticism that aircraft were landing only half-full and at other times with useless supplies, Milch himself ordered that some of the supply containers be opened and inspected before departure. To his horror he discovered many of the containers contained only fish meal, whereupon he returned them and asked the army to hang the victualing officer.[90] Pickert himself pointed out many years later that, "the fact that transport aircraft and para-dropped goods now and then contained foolish and unnecessary items is undisputed, but this was an exception which should not be overestimated."[91] Nevertheless, Milch wanted it stopped.

Milch did improve conditions at Gumrak. To enable his pilots to fly into Gumrak at night, he ordered lighting equipment, smoke pots, radio detecting equipment, and sent signals and air traffic experts into the cauldron.[92] Then he assured Paulus, Manstein and the High Command that on the night of the 18th, his aircraft would fly into Gumrak. He did not disappoint them. On 18 January, six He-111s and one FW-200, which alone carried six tons of supplies, landed at Gumrak and offloaded critical supplies.[93] Despite the improvements, the airfield was still more dangerous at night than during the day. Some 25 percent of the He-111s destined for the field either crashed or sustained damage during landing or takeoff, but the airlift into Gumrak continued steadily.[94] On the night of 21 and 22 January, the last night Gumrak was to be in German hands, twenty-one He-111s and four Ju-52s landed fully laden.[95]

The Red Army overran Gumrak on 22 January. This was a tremendous blow to the Sixth Army, which now was totally cut-off from the outside world except for airdrops, the Luftwaffe's final option.[96] The problem with airdrop missions is that they are inherently less efficient than airland missions. Airdrop loads require more time to pack, load, and rig the cargo for the airdrop. An airdrop mission, no matter how expert the crew, will seldom drop the supplies exactly where the customer (the Sixth Army in this case) demands, whereas an airland mission can put the cargo literally at the customer's feet. Airdrop missions result in more damaged cargo due to the impact velocity. It does not do the army any good to receive water they cannot drink, food they cannot eat, and bullets they cannot shoot. Airdrop has no provision for backhaul; there is no way to evacuate the wounded and sick.

When the fall of Gumrak appeared imminent, the Sixth Army had initiated a repair and construction program at Stalingradskiy, a lesser airfield further within the pocket.[97] The army reported it operational on the 22nd, but it was irrelevant. Stalingradskiy fell only hours after Gumrak.[98]

The Soviets continued their thrust through the Sixth Army, dividing it into two pockets; inexorably contracting the perimeters. All the Luftwaffe could provide now were just a few insignificant airdrops from the He-111s and Ju-52s. The situation was hopeless. On the 24th, Paulus, trapped in the southern pocket, sent an urgent request to the High Command: "Troops without ammunition and food. . . . Collapse inevitable. Army requests immediate permission to surrender in order to save the lives of remaining troops."[99] Hitler refused.

Many of the airdrop loads were irretrievable by the army; the emaciated troops simply lacked the strength to dig them out of the snow and there was no fuel left to transport them. To make matters worse, the transports were not able to clearly identify the locations of the German soldiers and scattered the canisters all over the pocket in an attempt to get *something* to them. Consequently, many of the supplies intended for the Sixth Army never made it to them. The supplies may have been dropped in the wrong area, lost in the city ruins or snow, blown out of reach, or simply landed in enemy sectors.[100] In a desperate attempt to improve the accuracy of the drops, on the evening of

25 January, Major Freudenfeld, the senior Luftwaffe signals officer, finished creating a drop site in the southern pocket. On 28 January, he created a drop zone in the northern pocket.

On 26 January, Paulus requested that the Luftwaffe drop only food. Ammunition was no longer needed since there were not enough guns.[101]

By dawn on the 31st of January, the southern pocket no longer existed; Paulus, having been promoted the day prior, capitulated, becoming the first German field marshal ever to be taken prisoner.[102] The Germans in the northern pocket continued fighting.

Army and air force leaders monitored their radios for further messages. The Luftwaffe refused to abandon the German soldiers. On 2 February, Milch ordered the aircraft to fly over the Stalingrad pocket and airdrop supplies to any clearly identifiable German troops. When the aircraft returned, Fiebig reported to Milch that it was hopeless. "The outline of the pocket can no longer be recognized. No artillery fire was seen. An enemy vehicle column with headlights blazing is advancing from the northwest into what was formerly the northern pocket. . . . The front of that column is almost at our former drop site."[103] These aircrew observations, combined with the lack of radio transmissions signified that the battle for Stalingrad was finally over.[104]

---

[1]Heinz Schröter, *Stalingrad*, trans. Constantine Fitzgibbon (New York: E. P. Dutton and Company, Inc., 1958), 261.

[2]Andrew Brookes, *Air War Over Russia* (Hersham: Ian Allan Publishing, 2003), 10; Murray, 77.

[3]Source information conflicts as to the actual number of troops trapped at Stalingrad. Estimates range from 200,000 to over 300,000. The number refers to the personnel with the Sixth Army when Soviet forces initially completed the encirclement. Bekker states the number was 250,000 (presumably based on the statement Richthofen made to Fiebig recounted later in this paper). Cajus Bekker, *The Luftwaffe War Diaries:*

The German Air Force in World War II, trans. and ed. Frank Ziegler (New York: Da Capo Press, 1994), 281; Edgar Howell, *The Soviet Partisan Movement 1941-1944, Department of the Army Pamphlet 20-244* (Washington: U.S. Government Printing Office, 1956), 41; Morzik cites the number to be "over 300,000," Morzik, 180; Pickert claims the number was 230,000. Wolfgang Pickert, "The Stalingrad Airlift: An Eyewitness Commentary", *Aerospace Historian* 18, December 1971, 183. Jukes says 330,000 men were encircled, Geoffrey Jukes, *Stalingrad: The Turning Point* (New York: Ballantine Books Inc., May 1972), 155; General Kurt Zeitzler stated after the war that estimates for the exact number of troops encircled varied from 216,000 to over 300,000. Kurt Zeitzler, "Stalingrad," in W. Kreipe, et al., *The Fatal Decision: Six Decisive Battles of the Second World War from the Viewpoint of the Vanquished,* ed. by William Richardson and Seymour Freidin, trans. Constantine Fitzgibbon (New York: William Sloane Associates, 1956), 160.

[4]Becker, 277-278; another source states that by mid-November, the Sixth Army held four-fifths of the city. N-17500.347-2, *The German Setback on the Southern Sector of the Eastern Front, 1942/43,* (Combined Arms Research Library, Fort Leavenworth, Kansas), 20-21.

[5]Becker, 278.

[6]Hayward provides an excellent and detailed analysis of the decision to execute the Stalingrad airlift. For additional information on the decision to execute the airlift, see Zeitzler, 160-174; Albert Speer, *Inside the Third Reich: Memoirs by Albert Speer,* trans. Richard and Clara Winston (New York: Macmillan Publishing Co,. Inc., 1970), 247-251.

[7]Joel Hayward, *Stopped at Stalingrad: The Luftwaffe and Hitler's Defeat in the East, 1942-1943* (University Press of Kansas: Lawrence, 1998), 244.

[8]Hayward, 245.

[9]Ibid., 245-246.

[10]Ibid., 246.

[11]BA/MA N671/9: *Richthofen TB*, 21 November 1942; quoted in Hayward, 226.

[12]Hayward's description of the conference is taken from Pickert's notes to his original diary entry (USAFHRA 168.7158-338: Pickert TB: "Aufzeichnungen aus meinem Tagebuch und von Besprechungen uber operative und taktische Gedanken und Massnahmen der 6. Armee"), from later correspondence between Picker and Hans Doerr (in USAFHRA K113.309-3 vol. 9), and from an essay written by Pickert to Hermann Plocher in 1956 (same source); quoted in Hayward, 236.

[13]Hayward, 236-237.

[14]Morzik, 188.

[15]Cooper, 252.

[16]Morzik, 189

[17]Ibid., 185; Hayward, 236; *The German Setback on the Southern Sector of the Eastern Front, 1942/43*, 16; Schröter, 163.

[18]James Corum, "The Development of Strategic Air War Concepts in Interwar Germany, 1919-1939," in *Distance Learning Version 3.0: Military Studies*, (Montgomery, Al.: Air Command and Staff College, 2000), 309.

[19]Cooper, 283.

[20]Morzik, 186-187.

[21]Ibid., 187.

[22]Ibid., 188.

[23]The distances between the airfields are inconsistent between sources. Cooper claims the distances stated in the text above, Cooper, 253; see also David Irving. *The Rise and Fall of the Luftwaffe: The Life of Luftwaffe Marshal Erhard Milch.* (Boston: Little, Brown and Company, 1973), 183. The scale provided on Bekker's rough sketch is difficult to interpolate with any precision, but it allows the reader to gauge the distance for himself. This author estimates the distance for Tatsinskaya to be 130 miles and that of Morozovskaya to be 115 miles; Vaughn and Donoho claim a distance of 140 miles. David K. Vaughn and James H. Donoho, "From Stalingrad to Khe Sanh: Factors in the Successful Use of Tactical Airlift to Support Isolated Land Battle Areas" (Maxwell Air Force Base: Air and Space Power Chronicles) [database on-line]; available from http://www.airpower.maxwell.af.mil/airchronicles/cc/vaughan.html; Internet; accessed 15 August 2003.

[24]Vaughan and Donoho.

[25]Asher Lee, *The German Air Force* (New York: Harper and Brothers Publishers, 1946), 287. The Ju-52 could carry between 1 and 2.5 tons of freight depending on the range of the flight. Since the operating bases were forced progressively further from their offload locations, airlift delivery tonnages steadily declined.

[26]Brookes says that there were *seven* airfields in the pocket, but only two of them were suitable as "logistical airheads," Brookes, 101.

[27]Hayward, 253.

[28]Vaughan and Donoho.

[29]Schröter, 169.

[30]Cooper, 255.

[31]Pickert, 184.

[32]Bekker, 288.

[33]Morzik, 190.

[34]Ibid.

[35]Hayward, 270-271.

[36]Bekker, 285.

[37]Hayward, 271.

[38]Bekker, 285; Hayward, 272; Morzik, 190; Hayward claims there were only 170 "airworthy" aircraft at Tatsinskaya, with 130 operable Ju-52s and 40 operable Ju-86s.

[39]Bekker, 285.

[40]Hayward, 271.

[41]USAFHRA 168.7158-335: *Fiebig TB*, 24 December 1942; quoted in Hayward, 272.

[42]Bekker, 286; Morzik, 190.

[43]Bekker, 286.

[44]Bekker states that nearly 60 aircraft, one third of the total, were left behind. Bekker, 287; Hayward cites BA/MA N6719: Richthofen TB, 24 December, 1942; USAFHRA 168.7158-335: *Fiebig TB*, 24 December 1942 and concludes that Fiebig lost 47 of 170 aircraft, for a total of 27 percent, Hayward, 272. Cooper's numbers list 109 salvaged Ju-52s and 16 Ju-86s, with 60 aircraft lost.

[45]Morzik, 190.

[46]Bekker, 286.

[47]Ibid., 287.

[48]Ibid., 287-288.

[49]Cooper, 253; Morzik, 191; Bekker's "sketch" depicts a distance closer to 200 miles. Bekker, 285.

[50]Morzik, 191.

[51]Hayward, 282. Hayward actually states the distances in kilometers: 330 kilometers from Novocherassk to Pitomnik, an increase of 130 kilometers over the previous routing from Morozovskaya. David Irving claims the distance was 220 miles. See Irving, 184.

[52]Bekker, 288.

[53]Cooper, 254.

[54]Ibid.

[55]Bekker 288.

[56]Hayward, 282, 291.

[57]*Battle for Stalingrad: The 1943 Soviet General Staff Study*, ed. Louis Rotundo (London: Pergamon-Brassey's International Defense Publisher's, Inc., 1989), 246.

[58]Brookes, 77; Murray, 156.

[59]Richard Suchenwirth, *Historical Turning Points in the German Air Force War Effort*, USAF Historical Studies, No. 189 (USAF Historical Division, Research Studies Institute, Air University, 1968), 108.

[60]*Stalingrad: An Eye Witness Account* (New York: Hutchinson and Company, 1945), 10.

[61]Morzik, 191; Hayward, 283.

[62]Morzik, 191-192.

[63]Bekker, 291.

[64]Ibid., 288.

[65]Ibid., 290. Bekker's actual words are, "A number of pilots were duly deceived and landed amongst the enemy."

[66]This number seems high. The numbers regarding specifics of the airlift are inconsistent between sources. According to "Zahlenangaben zur Luftversorgung Stalingrad," K113.309-3v9 USAFHRA; quoted in Thyssen, Mike, "A Desperate Struggle to Save a Condemned Army—A Critical Review of the Stalingrad Airlift," Research Paper, Air Command and Staff College, 1997; Pickert's records indicate that 24,910 wounded were evacuated. His own account many years after the war states that "some 30,000 wounded and sick soldiers were air-evacuated out of the pocket." Pickert, 184. Cooper claims the number is 34,000. Cooper, 255.

[67]Bekker, 288. Cooper does not mention the *Stuka*s and says the Me-109s were Bf-109s. Cooper, 254; Hayward, 288.

[68]Bekker, 288.

[69]Ibid., 291; Cooper, 254; Hayward, 288. Hayward states that "over half the planes flipped over and crashed when attempting to land."

[70]Cooper, 254.

[71]Pickert, 184.

[72]Hayward, 288.

[73]USAFHRA 168.7158-337: *KTB Sonderstab Milch*, 17 January 1943; quoted in Hayward, 288.

[74]USAFHRA 168.7158-335: *Fiebig TB*, 16 January 1942. See also Fiebig's complaint to Milch about the Sixth Army's refusal to expand Gumrak, USAFHRA 168.7158-337: *KTB Sonderstab Milch*, 17 January 1943; quoted in Hayward, 288.

[75]Williamson Murray, *Strategy for Defeat: the Luftwaffe, 1933-1945* (Maxwell Air Force Base: Air University Press, 1983), 9.

[76]USAFHRA 168.7158-337: *KTB Sonderstab Milch*, 15 January 1943; *KTB OKW*, vol. 3, 42 (15 January 1943); quoted in Hayward, 286.

[77]Cooper, 256.

[78]Hayward, 288.

[79]Hayward, 300.

[80]Cooper, 256.

[81]Hayward, 293.

[82]Morzik, 192.

[83]Pickert, 184.

[84]Bekker, 292.

[85]Bekker and Thyssen state that VIII Air Corps sent Maj Thiel. Hayward states that it was Milch. Hayward, 303.

[86]Hayward, 303.

[87]USAFHRA 168.7158-337: *KTB Sonderstab Milch*, 20 January 1943; quoted in Hayward, 303.

[88]Bekker includes an English translation of what this author believes is the entire report, although he does not mention the portion describing the crews unloading their own planes, Bekker, 292. Mike Thyssen paraphrases the original German account by Erich Thiel, "Meldung ueber Beschaffeneit des Platzes Gumrak und Ruedkgespraeche mit Generaloberst Paulus." P 1-10; quoted in Thyssen, 19.

[89]USAFHRA K113.309-3 vol.9: *Thiel, Major, Kommandeur iii./k.g. "Boelcke" Nr. 27, Gef.St., den 21.1.43. Betr. Meldung über Beschaffenheit des Platzes Gumrak (Kessel von Stalingrad) und Rücksprache mit Herrn Generaloberst Paulus*; quoted in Hayward, 303. Becker and Thyssen also include the Paulus quote, although each is slightly different than the one here, and slightly different from one another.

[90]David Irving, *The Rise and Fall of the Luftwaffe: The Life of Luftwaffe Marshal Erhard Milch* (Boston: Little, Brown and Company, 1973), 190.

[91]Pickert, 184.

[92]USAFHRA 168.7158-337: *KTB Sonderstab Milch*, 17 January 1943 (Milch-Fiebig conversation, 2033 hours); 17 January (Milch-Fiebig conversation, 2155 hours); 18 January (Milch Fiebig conversation, 0037 hours); USAFHRA 168.7158-335: *Fiebig TB*, 18 January 1943; quoted in Hayward, 297.

[93]Hayward, 297.

[94]For aircraft losses at Gumrak, see BA/MA RL 10/489: *Tagesabschlußmeldungen K.G. 55 vom 18 und 19.1.1943*; source cited in note 122 of chapter 6 (various entries during January); the diaries of Milch, Richthofen, and Fiebig; quoted in Hayward, 297.

[95]Bekker, 293.

[96]Morzik, 192.

[97]USAFHRA 168.7158-337: KTB Sonderstab Milch, 21 January 1943. For conditions at Stalingradskiy, see the source cited in note 38, esp. 9; quoted in Hayward, 302.

[98]BA/MA N671/10: *Richthofen TB*, 24 January 1943; BA/MA RL 10/489: *Tagesabschlußmeldung K.G. 55 vom 24.1.1943;* quoted in Hayward 302.

[99]Schröter, 225-26.

[100]Bekker, 294; Pickert, 184.

[101]Earl Ziemke and M. E. Bauer, *Moscow to Stalingrad: Decision in the East* (Washington D.C.: Center of Military History, United States Army, 1987), 78.

[102]Earl Ziemke, *Stalingrad to Berlin: The German Defeat in the East* (Washington D.C.: Center of Military History, United States Army, 1987), 79; Murray, 154.

[103]USAFHRA 168.7258-337: *KTB Sonderstab Milch*, 2 February 1943; quoted in Hayward, 310.

CHAPTER 4

STRATEGIC DILETTANTISM

> *Mein Führer*, Stalingrad has been the gravest crisis for the nation and armed forces so far. You must do something decisive to bring Germany out of this war. It is still not too late, and there are certainly many who think as I do. You must act now – act without ceremony, and above all act now.[1]
>
> Field Marshal Erhard Milch, *The Rise and Fall of the Luftwaffe*

If the Third Reich had a cohesive national grand strategy, it was specious at best. Hitler's ostensible capriciousness and masterful dilettantism with respect to the objectives and ultimatums he laid out before his armed forces resulted in their eventual inability to achieve success and in their ultimate destruction.[2] The obliteration of the Sixth Army was the first time that a Prussian or German field army had been encircled and annihilated since 1806.[3] In fact, never before in Germany's history had such a large number of troops come to such a gruesome demise.[4] Of the approximately three hundred sixty-four thousand soldiers[5] who approached Stalingrad in the summer of 1942, only ninety-one thousand men, half-frozen, weakened from starvation, and beginning to suffer from the throes of typhoid, marched out of the city and into Soviet prisoner of war camps. Along the way, the typhoid outbreak became an epidemic and killed about fifty thousand men. Many thousands died during the march to the camps in Siberia and Central Asia. Of the ninety-one thousand, only five thousand survived Soviet captivity, with the last prisoners returning to Germany as late as 1955.[6] The end of the fighting released no fewer than seven Soviet armies to attack elsewhere, exacerbating the increasing disparity in German-Axis weakness.[7]

The Soviets took advantage of the concentration of German military power at Stalingrad by switching over to the offensive along the entire front: in the north, the siege of Leningrad was broken on 18 January; along the Moscow Front, the Russians reclaimed Rzhev and Vyasma before the winter campaign ended; on the upper Don, they recaptured Voronezh prior to the surrender at Stalingrad. But the most rapid Soviet advances were in the South, where the Red Army bypassed the German forces at Stalingrad, establishing a new front along the Donetz and reclaiming Kharkov by mid February, and forcing the Germans to retreat from the Caucasus.[8]

Back in Germany, Soviet proclamations of a monumental victory forced the Nazi regime to reveal to the German people the loss of the entire Sixth Army.[9] Many Germans began to believe Germany would lose the war and Hitler himself became the target of widespread criticism for the first time.[10] After the war, Albert Speer, the Minister of Armaments and War Production for the Third Reich, wrote in his memoirs:

> Stalingrad had shaken us—not only the tragedy of the Sixth Army's soldiers, but even more, perhaps, the question of how such a disaster could have taken place under Hitler's orders. For hitherto there had always been a success to offset every setback; hitherto there had been a new triumph to compensate for all losses or at least make everyone forget them. Now for the first time we had suffered a defeat for which there was no compensation.[11]

The five months of fighting had destroyed 99 percent of Stalingrad.[12] A quick census before the battle revealed a population of more than nine hundred thousand residents, three hundred thousand of which were refugees.[13] At the end of the battle, Axis losses numbered one hundred fifty thousand. Russian losses were estimated at four to eight times greater, many of them civilian, although there are no firm estimates as to how many civilians lost their lives.[14] Most had perished in the opening days of the struggle or left for sanctuary in Siberia and Asia; no one knows how many had been killed, but estimates were astonishing.[15]

When Hitler learned of the outcome, he was furious over Paulus' decision to surrender, which upset him more than the destruction of the Sixth Army.[16] General Kurt Zeitzler, Chief of the General Staff, stated that the führer's only reaction, at least the only one he revealed to the officers around him, was to state that if he had expected that Paulus would surrender, he never would have promoted him to field marshal. He seemed completely unaffected by the bloody tragedy and the suffering of hundreds of thousands of his soldiers. Ever the optimist, he pushed the calamity out of his mind and started enthusiastically planning for the future, while casually telling Zeitzler that they would create a new Sixth Army. Hitler never admitted that he was to blame. Instead he insisted that he was always and invariably correct. Any misfortune, such as bad weather, was due to circumstances beyond his control. For several months Zeitzler struggled to ensure Hitler would learn the proper lessons and make the right decisions. Zeitzler himself admits that he failed. Consequently, he felt he would serve his country best by abdicating his position. When he informed Hitler, the latter was furious and replied roughly, "A general is not entitled to abandon his post."[17]

Hitler never publicly denounced Göring for the Stalingrad debacle.[18] In his memoirs, Field Marshall Erich von Manstein, commander of Army Group Don and responsible for ground operations in the southern Russia region including Stalingrad, observed the führer summoned him to his Supreme Headquarters on 5 February, just three days after the final struggles at Stalingrad. Manstein intended to ask the führer to step down as the commander in chief of the army and appoint an experienced and trustworthy general instead. Hitler may have been aware of Manstein's intentions because Hitler's opening words were:

> I alone bear the responsibility for Stalingrad! I could have perhaps put some blame on Göring by saying that he gave me an incorrect picture of the Luftwaffe's possibilities. But he has been appointed by me as my successor, and as such I cannot charge him with the responsibility for Stalingrad.[19]

The führer's frank admission disarmed the general; he decided not to tackle the issue, but instead, mildly suggested that Hitler appoint a competent and trustworthy chief of staff, whom Hitler could give authority over the other three service branches. Hitler responded that Göring would never respond to anyone's authority but his, that his own experiences with supreme armed forces commanders had always been disappointing, and that it was better if he remain in charge himself.[20]

Regardless of the apparent sincerity in Hitler's willingness to hold himself accountable and absolve Göring of any blame, the führer nevertheless lost more faith in the good judgment of his one-time first advisor.[21]

Göring's reaction was much more perplexing. When he learned of the Stalingrad debacle, he was overcome with grief and burst into fits of hysterical weeping; this was presumably an expression of sympathy for the thousands of Germans sacrificed and perhaps even an indication of guilt. It is possible that Göring's histrionics were based upon thoughts that were much more self-centered: Göring may have been more concerned with a loss of his prestige and fearful of further deflating his already tattering status with Hitler. Göring also persisted in his jealousy of Jeschonnek and went to great efforts to ensure that nothing would interfere with his sybaritic lifestyle. A more dangerous reaction, however, was Göring's complete sycophantic subservience to Hitler following the events at the city on the Volga. In a frantic effort to regain his führer's confidence, Göring became immediately acquiescent to Hitler's every whim. The Reichsmarschall's intellectual subservience was the last thing Germany's military needed. Contrary to what Hitler wanted--a bunch of automatons bowing to his every whim--what the Luftwaffe, Hitler, and the German people needed were advisors with the integrity and wherewithal to stand up to the führer and provide contrasting views. Perspective was needed. A yes-man like Göring lacked the strength and substance to do anything substantial. He could not save the already exhausted Luftwaffe from irrevocable and paralyzing exploitation.[22]

Milch returned to Germany along with General Hans Hube. Hitler summoned Hube to his private chamber and asked him suspiciously if the State Secretary had done everything in his power to ensure the success of the airlift. Hube replied, "All that and more!"[23] After Hube's departure Hitler met with Milch until 0130. Milch made it blatantly clear that if he had been Paulus, he would have disobeyed orders and commanded his army to break out of Stalingrad.[24] Hitler coldly replied that he would have no recourse but to lay Milch's head at his feet, and the field marshal, angry over the senseless waste of life, retorted, "*Mein Führer* – it would have been worth it! One field marshal sacrificed, to save three hundred thousand men!"[25]

Richthofen, like Milch, displayed resiliency, tenacity, and resourcefulness following the cessation of hostilities on 2 February. He immediately began rejuvenating his exhausted air fleet. He relieved his bombers of transport duties in preparation for close-support combat missions; he transferred other units to airfields further west where the operational rate increased due to improved weather, better facilities, shelter, and supplies. He also initiated the return of transport units to established and less crowded fields in the Crimea and southern Ukraine; he realigned, strengthened, and streamlined the chain of command; and he implemented an aircraft replacement and rehabilitation program. Steadily improving weather, February's major rehabilitation program, and the use of established airfields combined with Richthofen's other efforts as well as Milch's improvements raised operational ready rates and significantly improved combat effectiveness.[26]

During the seventy-two days and nights of the siege the Luftwaffe successfully delivered 8,350 tons of supplies into the cauldron, an average of 116 tons a day.[27] During this same period, the Luftwaffe had evacuated thirty thousand soldiers.[28] Only on three days (7, 21, and 31 December) did the transports manage to deliver three hundred tons into Stalingrad. The average amount was near one hundred tons per day and many days it was much less.[29] The German Air Force had endured the operation at a very high cost. Stalingrad proved to be the coup de grace for the Luftwaffe. Total aircraft losses from the Stalingrad tragedy were: 269 Ju-52s, 169 He-111s, 9 Fw-200s, 1 Ju-290, 5 He-177s, and 42-Ju 86s for a grand total of 495 aircraft. These losses were the equivalent of five flying wings or an entire air corps.[30] Certainly the losses in materiel were formidable, but they were overcome. However, the loss in manpower was not. Göring was certainly referring to his aircrews and training schools as well as He-111s when he told interrogators after the war, "I built the Luftwaffe as the finest bomber fleet, only to see it wasted on Stalingrad. My beautiful bomber fleet was used up in transporting munitions and supplies to the army."[31] It was to prove to be the death of the Luftwaffe.

German transport aircraft took on a renewed importance to Hitler, albeit briefly. In the aftermath of the Stalingrad catastrophe, Hitler demanded of Milch, "I want transporters, transporters, and more transporters!"[32] He even suggested a primitive

aircraft with the capability of delivering four tons of cargo on rough, unprepared surfaces.

However, the führer's rash demand for increased production of airlift aircraft to support

his armies never materialized. His desultory tendencies and ephemeral interest in

transports revealed themselves when the Wehrmacht began evacuating the Caucuses

towards the Crimea, because Hitler now asked for Ju-52 seaplanes, too.[33] Germany never

had the luxury of increasing airlift production as more pressing concerns demanded

increases in other types of aircraft to stem the tide of the Allied forces. The atrophying of

German aircraft, especially the airlift assets, was approaching the breaking point. On 5

March, Milch was dining alone with Hitler. When discussing the planned spring

offensive in the east,[34] Milch warned Hitler that German forces were presently too weak

and their transports were inadequate over such great distances.[35]

The Luftwaffe's losses were not just a localized phenomenon around Stalingrad.

To achieve the concentration of airlift necessary to save Paulus's army, the German Air

Force was forced to weaken its concentration elsewhere, and it was never able to restore

its erstwhile strength.[36] In December 1942, 36 percent of the operational first-line aircraft

along the Eastern Front were concentrated in the Don-Donetz sector. In the desperate

attempt to save the Sixth Army by early February, this number had jumped to 950 of

approximately 1,800 total, or 53 percent.[37] The aircraft left on the other sectors were only

useful for reconnaissance and other noncombatant duties. Consequently, the battle of

Stalingrad left the German Air Force in other Russian sectors unable to cope with the

duties confronting them, and without air superiority.[38] The reallocation of bomber units to

air transport duties exacerbated Luftwaffe disorganization and manifested itself in extensive decentralization of operational control. This highlighted itself when Fiebig's VIII Air Corps was stripped of its combat duties and assumed responsibility for the Stalingrad airlift. From this point forward, the Luftwaffe took on the role of an ineffective defensive force, attempting to neutralize Soviet advances as a salve applied to the wounds the Red Army inflicted on the unfortunate Germans.[39]

While the German Air Force was inexorably hemorrhaging, the Soviet Air Force was displaying remarkable resiliency and strength. Whereas a depleted and exhausted Luftwaffe was forced to overextend itself on three fronts, the Red Air Force had the benefit of confronting their German nemesis with ever increasing aircraft. Matthew Cooper points out that in mid-January 1943, the severe losses had reduced the combat strength of the Luftwaffe in the East to just 1,700 aircraft. This was only 60 percent of the force that had commenced operations only six months earlier, and only 20 percent of the force the Soviets were currently throwing at them. By 1943, the Soviets possessed at least 5,000 front-line aircraft, of which two-thirds were in the South. These airplanes were better suited than those of the Luftwaffe to operate in the cold climate. Poor flying conditions, low serviceability, and a loss of forward airfields for fighter units exacerbated the paucity of German aircraft, resulting in their inability to obtain air superiority even in the Don-Rostov area, where the Luftwaffe had concentrated 52 percent of their strength in the East.[40]

Regardless of Hitler's transitory outburst demanding an increase in transports, other priorities took precedence. The führer's aircraft requirements were capricious expedients, which varied with the tide of battle.[41] Additionally, the resurgent Soviet Air Force, teamed with Allied successes in the Mediterranean and the Combined Bomber

Offensive forced the Germans' hand in terms of aircraft production. This was the most important decision the Luftwaffe faced during the Stalingrad crisis (except, of course, committing to the airlift in the first place).[42] Resources were scarce. Countering the growing threat of a multifront war against an enemy with increasing numerical and technological superiority dictated that the Germans multiply their fighter output at the expense of all other platforms. The transports did not stand a chance.

When the war began, the Luftwaffe possessed 550 Ju-52s, which was the only aircraft capable of fulfilling transport duties at that time; it was the backbone of the transport units.[43] In March 1942, Milch recognized the need for a tremendous increase in single-engine fighters and proposed a new production program to increase fighter output to one thousand per month. However, both Göring and Hitler insisted that an increase in defensive capability must not come at the expense of offensive capability, with bombers retaining priority. Hitler further required that the output of transport aircraft be raised to four hundred per month, to include a large number of troop-carrying types.[44] A few months later, Milch proposed a second program that, in addition to tripling the fighter output of his first program by the summer of 1944, would also produce five hundred training aircraft per month.[45]

In spite of Milch's proposals to increase output of transports and trainers, there was no noticeable change. Cooper says that by the end of 1942, the number of Ju-52s had only risen to eight hundred, which was insufficient to meet the Luftwaffe's deepening involvement in the Wehrmacht's ever increasing commitments. Consequently, many other aircraft were adapted for transport roles due to the serious shortage of Ju-52s. By the end of 1942, there were twice as many Ju-52s lost as there were being produced.[46]

When the war started, the Training Command possessed two-thirds of the available Ju-52s. These were used in bomber, blind-flying, instrument flying, and bomber-observer schools. During a major operation, there was no alternative other than to reallocate them and their crews to the front, at least temporarily. As an example, 380 Ju-52s were taken from their training bases and used in one Western campaign for ten weeks, where 150 were destroyed. In December 1941, Hitler ordered the establishment of five new transport groups for the Eastern Front. The aircraft and personnel were taken from the training schools, nearly stripping them of their equipment, instructors, and advanced students.[47]

Major General Paul Deichmann, who was in charge of Luftwaffe Command Four at war's end, stated after the war:

> It is a well-known fact that the practice of requisitioning Ju-52s from the training schools continued unabated and, in fact, became more and more common as the war progressed. As a result, of course, the schools were simply unable to fulfill their mission of providing trained replacement personnel for the bomber and long-range reconnaissance forces.[48]

The exceptional scale of transportation operations created a serious fuel shortage, aggravating an already critical shortfall in the training program. These shortages would have lasting consequences. They began in August 1942 and became progressively worse as the demands of the Stalingrad campaign multiplied. The Germans took drastic measures to minimize the effects, in particular a restriction on all flying behind the fronts. This limitation crippled an already decimated pilot training program and its repercussions would further attenuate the efficacy of the Luftwaffe.[49] In some theaters the German Air Force was forced to prohibit all flying except for operational flights absolutely vital to the war effort. The shortage continued to make itself felt until the spring of 1943, and even until June, the stringent and deleterious rationing continued. Only the development of the synthetic oil industry relieved the pressure, restabilizing the situation by midsummer 1943. The crippling fuel deficit compromised long-term plans, and the Luftwaffe used 1943 for rebuilding instead of seizing the initiative.[50]

While the fuel shortage was one problem, a modernized airlift fleet was another. Germany needed an updated aircraft to replace the Ju-52. The transport fleet was reeling from a crushing blow, and although it eventually replaced its losses, this was accomplished mainly by usurping foreign aircraft--for example, Italian aircraft after Italy's surrender in September 1943. The German aircraft manufacturing program had to reevaluate its desire to replace the Ju-52 with the Ju-352, settling instead for operational aircraft since they were more economical to build and needed immediately due to the exigencies of war.[51]

The destruction of the Sixth Army did not allow the VIII Air Corps even a brief respite for the refitting and reconstituting they so desperately needed. Instead they were released to provide air supply and combat support to the Seventeenth Army in the Caucasus's Kuban bridgehead.[52] The Kuban River flows through the Taman peninsula, which separates the Black Sea from the Sea of Azov. German Army reverses farther north along the front precipitated a necessary withdrawal from the Caucasus in early 1943. Hitler insisted on maintaining his forces on both sides of the Kuban as a starting point for a renewed offensive into the Caucasian oil fields later in 1943. Additionally, its loss might have allowed a Soviet invasion into the Crimea, thereby cutting off the entire German southern flank and the crucial Rumanian oil fields. So on 23 January 1943, Hitler ordered a series of defensive positions on both banks of the river to establish a bridgehead. Since the Germans maintained a limited naval presence in the Black Sea, the responsibility for sustaining the troops on the Kuban bridgehead fell to the Luftwaffe.[53]

The transporters' mission was to deliver ammunition, gasoline, and food to the Kuban bridgehead. All available space would be used for wounded and other specifically designated personnel on the return leg. No daily quota was established; the operation would last as long as the ground situation warranted.[54]

On 4 February, just two days after the last airlift missions to Stalingrad, FW 200 Condors, a four-engine, civilian transport converted to a military role, flew the first transport missions. By the middle of February, 180 Ju-52s, together with a small group of gliders, had replaced the Condors, which were desperately needed in the West.[55] By the end of March, the Germans had reestablished communications over the Straits of Kerch, which separates the Kuban from the Crimea. During the fifty days of the airlift, the Luftwaffe averaged 182 tons a day to the Seventeenth Army, 5,148 tons in all. This number is quite an improvement over the daily average managed for the Sixth Army by a much larger force only several months earlier. Again, the Luftwaffe fulfilled Hitler's demands and supplied an army in the field, but the weather and tactical situations were more favorable for the Kuban airlift.[56]

Despite the success of the airlift, the overall outcome for the remainder of the German Air Force assets was much less reassuring. Hitler's insistence on maintaining the Kuban tied up precious ground and air forces urgently needed elsewhere and exacerbated the attrition toll.[57] While the Germans were continuing their life-or-death struggle along the Eastern Front, they were simultaneously fighting additional campaigns in the Mediterranean and North Africa.

Although a detailed account of the airlift operations in the Mediterranean is beyond the purview of this paper, the endeavors represent the ubiquitous strains and attrition placed upon the Luftwaffe as a whole, and the transport forces in particular, and so deserve a brief synopsis.

The Allies had handed the Germans a crisis in North Africa before the Sixth Army's progress at Stalingrad deteriorated. The British victory at El Alamein in November 1942, coupled with the Allied landings of Operation Torch, forced the Luftwaffe to abandon over two hundred aircraft,[58] while almost simultaneously dispatching 320 Ju-52s to the Mediterranean during November. In November and December, the transports delivered 41,768 personnel, 8,614.8 tons of equipment and supplies, and 1,472.8 tons of fuel.[59] If such an airlift can be declared a victory, it was a Pyrrhic victory: the Luftwaffe lost no less than 154 Ju-52s by the end of January. Combined with the losses at Stalingrad, the Germans lost 659 transport aircraft (56 percent of their transports as of November 10) before Paulus' surrender.[60] Comparing production to destruction, in the first half of 1942 Germany produced only 235 new Ju-52s to replace the 516 that had been sacrificed.[61] The Luftwaffe certainly could not afford to continue fighting a protracted war.

As Allied operations progressed in North Africa, Tunisia became the target area. Overwhelming Allied air and naval superiority, as well as "Ultra" information provided allied commanders with the means to cut the German lines of communication, making airlift the only option.[62] The Luftwaffe reallocated 250 various transport aircraft, almost all of them Ju-52s, to supply the Wehrmacht in an effort to meet the impending Allied advances.[63] Although this airlift proved to be very similar to Stalingrad, it was longer and even more costly for the Luftwaffe.[64] Another significant factor was that these aircraft were wasted on an operation with little strategic value compared to the Don-Donetz region,[65] while Paulus and his men were freezing, starving and inexorably bleeding to death due to the lack of operational airlift.

And so it was to continue along these same lines for the remainder of the war. The Luftwaffe simply could not meet all the tasks required of it.[66] The German Air Force was assigned one *fait accompli* after another. The Luftwaffe was Hitler's fire hose, used carelessly in increasingly vain attempts to squelch the conflagrations the führer started when he decided on war. The problem was that with the Luftwaffe committed on so many fronts, he had no hope of success. The Luftwaffe was called upon time and again to support the desperate situations shaping the ground campaigns, as if Germany's Air Force was a fungible asset that could be replaced as easily as changing a roll of paper towels. The reality, however, was that the true life blood of the Luftwaffe was the human. Milch captured the essence of this during a Central Planning session. During the meeting Speer was lamenting on the scarcity of materials and fuel when Milch cut him off stating, "The only raw material which cannot be restored in the foreseeable future is human blood."[67]

By 1942, the training program was extinct in terms of its ability to produce the caliber of pilots demanded to satisfactorily engage their adversaries.[68] The loss of pilots and skilled aircrews was probably the decisive factor in the collapse of the Luftwaffe as an efficacious fighting machine.[69] Manpower was the one resource the Luftwaffe could not effectively replace. The cumulative attrition over the battlefields, closing of the training schools, and severe limitations in fuel due largely to the airlift operations meant

74

that the German Air Force was suffering from a degenerative disease from which it would find no cure. The average life expectancy of a German line pilot over the course of the war was between eight and thirty days, and their attrition rate was most likely well into the 90th percentile, Murray concludes. The statistics for the pilots of other aircraft could not have been much better.[70] The loss of skilled, experienced pilots forced the Germans to shorten training programs in order to fill cockpits with increasingly less skilled pilots. These new pilots were, in turn, lost at a faster rate, which forced the training programs to produce pilots even faster.[71] When factoring in the loss of the instructor crews from the training schools, it is clear that the German Air Force was caught in a vicious cycle from which there was no escape, and this cycle continued until the utter collapse of the Luftwaffe shortly before the disintegration of the Third Reich.

---

[1]David Irving, *The Rise and Fall of the Luftwaffe: The Life of Field Marshal Erhard Milch* (Toronto: Little, Brown and Company Limited, 1973), 202. Milch made this statement to Hitler in March 1943, emphasizing the precariousness of Germany's position following the disaster at Stalingrad.

[2]While in Spandau prison after the war, Speer wrote in his diary, "Someone ought to write on Hitler's dilettantism someday. He had the ignorance, the curiosity, the enthusiasm, and the temerity of a born dilettante; and along with that, inspiration, imagination, lack of bias. In short, if I had to find a phrase to fit him, to sum him up aptly and succinctly, I would say that he was a genius of dilettantism." Albert Speer, *Spandau: The Secret Diaries*, trans. Richard and Clara Winston (New York: Macmillan Publishing Company, Inc., 1976), 347.

[3]Hermann Plocher, *The German Air Force Versus Russia, 1942,* USAF Historical Studies, No. 154 (USAF Historical Division, Aerospace Studies Institute, Air University, June), xiii.

[4]Siegfried Westphal, "Between the Acts," in W. Kreipe, et al., *The Fatal Decision: Six Decisive Battles of the Second World War from the Viewpoint of the Vanquished,* ed. William Richardson and Seymour Freidin, trans. Constantine Fitzgibbon (New York: William Sloane Associates, 1956), 190.

[5]Heinz Schröter, *Stalingrad*, trans. Constantine Fitzgibbon (New York: E. P. Dutton and Company, Inc., 1958), 261.

[6]Geoffrey Jukes, *Stalingrad: The Turning Point* (New York: Ballantine Books Inc., May 1972), 155; Earl Ziemke, *Stalingrad to Berlin: The German Defeat in the East* (Washington D.C.: Center of Military History, United States Army, 1987), 79; David Irving places the number of soldiers entering Soviet captivity at 108,000. Irving, 197.

[7]David Irving, *The Rise and Fall of the Luftwaffe: The Life of Field Marshal Erhard Milch* (Boston: Little, Brown and Company, 1974), 197.

[8]British Air Ministry, *The Rise and Fall of the German Air Force: 1933-1945* (New York: St Martin's Press, 1983), 224. Cited hereafter as British Air Ministry.

[9]William Craig, *Enemy at the Gates* (New York: Reader's Digest Press, 1973), 384.

[10]Ian Kershaw, *Hitler 1936-45: Nemesis*, (New York: W. W. Norton and Company, 2000), 551.

[11]Albert Speer, *Inside the Third Reich: Memoirs by Albert Speer*, trans. Richard and Clara Winston (New York: Macmillan Publishing Company Inc., 1970), 254.

[12]Craig, 385.

[13]Roger Spiller, *Sharp Corners: Urban Operations at Century's End* (Fort Leavenworth, KS: Combat Studies Institute, 2001), 50-55.

[14]G. F. Krivosheev, ed., *Soviet Casualties and Combat Losses in the Twentieth Century*, (London: Greenhill Books, 1997), 123-25; quoted in S. J. Lewis, "The Battle of Stalingrad" in *Block by Block: The Challenge of Urban Operations*, ed., William G. Robertson and Lawrence A. Yates (Fort Leavenworth: US Army Command and General Staff College Press, 2003), 53.

[15]Craig, 385.

[16]Ziemke, 78-79.

[17]Zeitzler points out that the Sixth Army could never be recreated. "It had died at Stalingrad. With it had died a large part of the confidence which the German Army had hitherto felt in its Supreme Commander; or, to be more exact, he had killed it by his own obstinacy." Kurt Zeitzler, "Stalingrad," in W. Kreipe, et al., *The Fatal Decision: Six Decisive Battles of the Second World War from the Viewpoint of the Vanquished*, ed. William Richardson and Seymour Freidin, trans. Constantine Fitzgibbon (New York: William Sloane Associates, 1956), 188-89. Apparently, Zeitzler wasn't aware of Hitler's comments to Manstein on 5 Feb, where he (Hitler) accepted full responsibility for the

events at Stalingrad, indicating Hitler's cunning and uncanny ability to isolate his generals while simultaneously manipulating them.

[18]Richard Suchenwirth, *Historical Turning Points in the German Air Force War Effort*, USAF Historical Studies, No. 189 (USAF Historical Division, Research Studies Institute, Air University, 1968), 108.

[19]Erich von Manstein, *Lost Victories*, trans. and ed. Anthony G. Powell (Novato, Ca.: Presidio Press, 1994), 365.

[20]Joel Hayward, *Stopped at Stalingrad: The Luftwaffe and Hitler's Defeat in the East, 1942-1943* (University Press of Kansas: Lawrence, 1998), 321.

[21]Suchenwirth, 108.

[22]Ibid.

[23]Field Marshal Erhard Milch, *Memoirs*, unpublished, 1946-47; quoted in Irving, 196.

[24]Milch Diary, 28 September 1947; quoted in Irving, 196.

[25]Irving, 196.

[26]Cooper, 155; Hayward, 327; British Air Ministry, 227-28.

[27]Irving, 197; Hayward cites the daily average as 117.6. Hayward, 310; Cooper says the daily average was 94 tons. Cooper, 294; Morzik also claims an average of 94 tons per day. Fritz Morzik, *German Air Force Airlift Operations*, USAF Historical Studies, No. 176 (USAF Historical Division, Research Studies Institute, Air University, 1961), 214. Whatever the actual number, it was still far below the amount required to sustain the Sixth Army.

[28]"According to Pickert's figures for 25 November 1942 to 11 January 1943, air transport groups evacuated 24,190 wounded troops, a daily average of 519 (USAFHRA 168.7158-338: 'Pickert airlift statistics');" quoted in Hayward, 310. Unfortunately, Pickert's figures do not include the remaining days of January 1943. In his postwar article, "The Stalingrad Airlift: An Eyewitness Commentary, Pickert wrote that "some 30,000 wounded and sick soldiers were air-evacuated out of the pocket," Wolfgang Pickert, "The Stalingrad Airlift: An Eyewitness Commentary," *Aerospace Historian* 18 (December 1971): 184.

[29]"Luftversorgung der 6. Armee vom 24.11.12. bis 3.2.43.," NARS T-321/18/4758846; quoted in Williamson Murray, *Strategy for Defeat: The Luftwaffe, 1933-1945* (Maxwell Air Force Base, AL: Air University Press, 1983), 154; Irving puts the total aircraft at "488 aircraft destroyed, missing, written off, and about 1,000 airmen," 197, and he includes the following in his endnotes: "In comparison with sorties flown,

loss rates were: Ju 52: 10 percent; He 111: 5.5 percent; Ju 86: 21 percent; He 177: 26 percent. The British Bomber Command operated on the assumption that no air force could maintain flying operations in the face of a sustained loss-rate exceeding 5 percent, which again testifies to the courage of the Luftwaffe aircrews." Irving, 396; Plocher also lists the total number of aircraft and personnel lost as 488 and 1,000 respectively. Plocher credits the numbers for aircraft losses to an entry in the Milch diary dated 2 February 1943, as cited by von Rohden, p 140. The numbers include those sustained by the 4th Bomber Wing. Plocher, 354.

[30]Murray, 155.

[31]N-9618, Interrogation *of Reich Marshall Hermann Goering, 10 May 1945, 1700 to 1900 hours* (Combined Arms Research Library, Fort Leavenworth, KS), 4, 10; N-10007-3, *Headquarters Air P/W Interrogation Detachment Military Intelligence Service: Hermann Goering, 1 June 1945* (Combined Arms Research Library, Fort Leavenworth, KS), 10.

[32]Milch Diary, 4 Feb 1943; quoted in Irving, 198.

[33]Milch Diary, 17 Feb, 1943; quoted in Irving, 198.

[34]This would be "their last big ground offensive on the Eastern Front, the attack on Kursk, which, if successful, would eliminate the main Russian salient on the central front, split the Red Army and so discourage it from any further thrust westwards in the Smolensk province farther north." Lee, 160.

[35]Irving, 202.

[36]Lee, 155.

[37]British Air Ministry, 224-25; The Luftwaffe on the Don front was increased by some 550 operational aircraft, of which some 350 were operational types and the remainder were transports, resulting in the concentration of 1,000 aircraft. Asher Lee, *The German Air Force* (New York: Harper and Brothers Publishers, 1946), 154.

[38]British Air Ministry, 225.

[39]Lee, 155; British Air Ministry, 208.

[40]Matthew Cooper, *The German Air Force, 1933-1945: An Anatomy of Failure* (New York: Jane's Publishing Incorporated, 1981), 257.

[41]Irving, 198.

[42]British Air Ministry, 205.

[43]Cooper, 283.

[44]British Air Ministry, 207.

[45]Ibid.

[46]Cooper, 283.

[47]Ibid.

[48]Andreas Nielsen, *The German Air Force General Staff,* USAF Historical Study, No. 173 (New York: Arno Press, 1959), 160.

[49]British Air Ministry, 218.

[50]Ibid.

[51]Ibid.

[52]Kampfgeschwader 4, Geschichte des Kampfgeschwaders "General Wever" Nr. 4, III. /K. G.4 im Osten, 1943 (Bomber Wing 4, History of the Bomber Wing "General Wever" No. 4, 3rd Group, Bomber Wing 4 in the East, 1943), Karlsruhe Document Collection; quoted in Plocher, 17; Cooper, 293.

[53]Andrew Brookes, *Air War Over Russia* (Hersham: Ian Allan Publishing, 2003), 10; Murray, 156.

[54]Morzik, 205.

[55]Cooper, 293.

[56]Ibid., 294.

[57]See Hermann Plocher, *The German Air Force Versus Russia, 1943,* USAF Historical Studies, No. 155 (USAF Historical Division, Aerospace Studies Institute, Air University, 1967), Chapter 2.

[58]Lee, 134.

[59]Murray, 160.

[60]"Based on the quartermaster general's loss tables for November-December 1942, and January 1943, BA/MA, RL 2 III/1184, 1185, Genst. Gen. Qu. (6.Abt), 'Flugzeugunfälle und Verluste bei den fliegenden Verbänden.'"; quoted in Murray, 160.

[61]Brookes, 77.

[62]Murray, 162-63; British Air Ministry, 154.

[63]Hayward, 316; British Air Ministry lists the breakout of transports moved to support operations in Tunisia as 20 Me-323s and 320 Ju-52s. These aircraft came at the expense of the Russian front and the training program. British Air Ministry, 159.

[64]Murray, 162.

[65]Hayward, 316.

[66]Cooper, 295.

[67]Central Planning, 12 Feb 1943 (Milch Diary, 47, p. 9408); quoted in Irving, 199.

[68]Suchenwirth, 27.

[69]Murray, 303; Nielsen, 173-74.

[70]Murray, 303. While Murray does not specifically mention transports, their losses were probably as high as the other types.

[71]Ibid., 312.

CHAPTER 5

CONCLUSION

.

> In starting and waging a war, it is not right that matters, but victory.[1]

> Adolf Hitler, *Air War Over Russia*

It can be argued that Hitler's policy of never giving back land that had already been won was the major cause for the Sixth Army's destruction at Stalingrad. But Hitler's obstinacy is a superficial symptom of a much greater illness that was systemic to Germany's strategic-level military leaders. Hitler seemed to suffer from an almost pathological blindness towards his mistakes. If he was aware of his errors, he did not learn from them. In the Stalingrad debacle, Hitler not only failed to learn from his mistakes, he even learned some negative lessons.[2] Senior Luftwaffe leadership was no better. The events at Stalingrad failed to spark within them any action to overcome their jealousies and forge a spirit of cooperation. There is no evidence that the Luftwaffe High Command made any changes in its strategic planning after the tragedy at Stalingrad.[3] This is especially sobering considering Rommel's surrender in Africa and the Allies' successful landing there.

One can only speculate as to how Hitler would have reacted if he had an accurate and complete understanding of the pernicious ground situation that confronted Paulus and his men. What would Hitler's response have been if Luftwaffe leadership had outlined the number of aircraft required to guarantee resupply for an entire army? Had anyone briefed Hitler of the extreme measures the Luftwaffe had to take to meet the lift requirements necessary for the soldiers at Demyansk and Kholm, would he have allowed the pillaging of the training schools' aircraft and crews? While there is no evidence to indicate Hitler was aware of the previous sacrifices, if the Germans had recognized the inherent capabilities of their airlift forces, and created an air transport fleet, it is difficult to conclude that Hitler would not have known the cost-benefit analysis of undertaking large-scale airlift operations in a hostile environment. This notional air fleet staff would have had facts and figures at their fingertips, and used the evidence to urge Göring, if not Hitler himself, not to rely on the airlift. If the führer understood that he did not have enough aircraft to support an airlift for two hundred fifty thousand men, especially considering the expected attrition due to the Russian winter, operations from primitive airstrips, and battle losses, would he still have ordered his men to stand resolutely and fight? Or would Hitler have authorized a breakout from the siege while Paulus and his men still (arguably) possessed the equipment, ammunition, fuel, and strength to attempt one? With the benefit of hindsight, it seems that the obvious choice should have been a breakout and that this order should have been given once it became apparent that it would be impossible to supply the Sixth Army, either by the organic German logistical system or by the Luftwaffe's airlift aircraft. The success of such a breakout is another cause for speculation and debate, but it certainly would have provided the Sixth Army with the possibility for survival, unlike the agonizing suffering and ignominious defeat they suffered while standing firm.

General Zeitzler claimed that the destruction of the Sixth Army at Stalingrad was the turning point of the entire war.[4] While *all* historians will probably never unequivocally agree with his assessment, a conclusion that should stir much less controversy is that it was Germany's military leaders' haphazard and almost erratic military strategies, policies, and directives that led to the downfall of Germany as a whole, and the Luftwaffe in particular. Suchenwirth concluded that it was the air supply operations, not one of which was imperative for military necessity, which destroyed the Luftwaffe.[5] Hitler's unwillingness to act on the advice of his staff, his stubbornness, dilettantism, and lack of a cohesive strategy offers a remarkable illustration of incompetence. His astonishing megalomania and grandiose schemes for world domination led to his spreading his forces beyond the breaking point. His desire to conquer and control *everything* inexorably led to his inability to control *anything*.

<u>Relevance to the Contemporary Environment</u>

Worldwide commitments are stretching the United States strategic airlift capabilities to dangerous levels. General John W. Handy, commander of United States Transportation Command (TRANSCOM) and Air Mobility Command, said on 25 June 2003 that despite six meetings with General Tommy R. Franks, the US Central Command and overall architect of the most recent Gulf War, he could not meet the needs of the war and other theaters due to insufficient lift, and he and Franks had to "negotiate" the use of TRANSCOM assets.[6] Although impending strategic lift deficiencies raise concerns over US abilities to sustain long-term commitments in multiple theaters, the bigger concern is the proliferation of ground commitments demanding the stretching of strategic airlift. In Germany's case, it was the Wehrmacht's ubiquitous ground commitments that precipitated the Luftwaffe's demise.

The recent explosion of terrorist activities threatens US strategic interests abroad and at home. The continental US faces the most certain foreign threat since the war of 1812.[7] Consequently, as of summer 2003, a higher percentage of the total army is

83

committed to active combat operations than during any time since World War II. Of the 495,000 troops in the U.S. Army, 375,000 are currently deployed around the world.[8] In late winter of 2003, Special Operations Command was deployed in sixty-five countries.[9] While there are arguments that the military is too small to meet global commitments,[10] a more useful and infinitely more difficult question to answer is: are US commitments too large? The long-range outlook for the effectiveness of the Army may be questionable.

Unfortunately, the frenetic pace of military operations may be damaging the cultivation of the critical skills crucial to successful operations in today's military environment. In an age of increased commitments and more stringent military spending, Americans need qualified, educated, and professional officers more than ever. However, the military appears reticent at best, and loathe at worst, to ensure their educational institutions are fostering the critical thinking skills and intellectual diversity officers in the twenty-first century need. Historian Williamson Murray describes the insidious weakening of the military's educational institutions:

> Teaching duty on the faculties of professional military schools is still not "career enhancing"; the navy still refuses to send a substantial number of its best officers to any school of professional military education; the Army War College, despite an impressive faculty, is an institution where war rarely appears in the curriculum; the army has turned one of its few truly innovative educational experiments of the 1980s, SAMS, into a humdrum planning exercise; the Air War College, after a short period of professional military education, has returned to the golf course; and finally, the National War College remains buried within the army's budget, where it simply fails to get the support it needs.[11]

Understanding the political context of war and a sincere appreciation of the enemy's way of thinking are vital for success in the current operating environment. Officers require knowledge of foreign languages, cultures, religious beliefs, and most important of all, history.[12] Army officials recently revealed a disturbing development: the in-residence portion of the primary training program for army company grade officers, Combined Arms and Services Staff School (CAS3), is terminating.[13] Rather than spending six weeks attending an institutionalized training program, where army officers are afforded the opportunity to interact with officers from other branches, the new plan calls for them to spend an additional two weeks in their respective captain's career courses. This lack of interaction with officers from different branches will prevent the cross-fertilization of ideas, perspective, and experience from other soldiers. Is this weakening in one of the Army's foundational training programs a coincidence, or is this a consequence of an over-commitment of forces due to the exigencies of US compliance with President George W. Bush's *National Security Strategy of the United States of America,* September 2002, and dedication to the global war on terrorism? Is the Army shortchanging these officers and tacitly negating its ability to underwrite America's political, economic, social and military objectives? Is there another mechanism to replace this valuable training? So far this is an isolated incident, and not intended to sound alarmist. However, it is worth noting that this may be an indicator of future developments, especially if the military maintains its current operational tempo with its current force structure.[14] If so, this may be the first step towards a gradual, eroding decline in the expertise of America's military that will have resonating implications in the Army's ability to employ a credible, efficacious fighting force.

The objective of this discussion is not to paint a cynical view of the George W. Bush Administration's policies or the importance of meeting terrorism head-on. Certainly, such resolve is essential to the long-term health, if not existence, of this nation. Rather, the real impetus behind the debate is how much is too much? How long is too long? How far is too far? Will the US military reach the breaking point? The answers to these questions do not yet exist, but reflection on their consequences should not be avoided.

To better understand the effect current global commitments may have on the Army, one can look at the Army Reserve and National Guard. Over two hundred thousand of these troops have been called up to support operations in Iraq, Afghanistan, and the US. Some of them have been deployed for more than a year, earning a fraction of their civilian pay and raising concerns that the hardships on them and their families may hurt Guard recruiting.[15] The regular Army may not fare much better. An all volunteer force that is perpetually separated from family and friends, in austere environments, under hostile conditions gives little incentive for reenlistment. These comments are not intended to disparage the loyalty and tenacity of the American soldier, but in the era of an extended duration war with an all-volunteer service, one has to wonder what the future holds in store. This comparison may be criticized as "apples and oranges" since Luftwaffe casualties and attrition were exponentially higher than those suffered by US forces today. However, the long-term effects are the same. Administrative attrition and declining combat capabilities due to training negligence can be just as severe as the Luftwaffe's mistakes and their unfortunate results.

US policies and efforts to contain worldwide terrorism have spread the military, and particularly the Army, very thin. With desires to maintain, if not increase, global influence and stability, is the US heading toward military collapse? The Luftwaffe was forced to close its training schools to meet the danger of a multifront war. By spreading it forces to support situations on the ground, the Luftwaffe was unable to concentrate its firepower, and was slowly ground down. Army losses today are nowhere close to what the Luftwaffe suffered, but today's military faces an administrative attrition: elective separation. Additionally, the Army may be fielding less than fully trained officers. The Army is shortening its officer training program, ostensibly to increase the throughput of officers to the field. In an effort to control *everything*, is America inexorably increasing her inability to control *anything*?

An additional parallel one can draw from Germany's experience on the Volga is the United States' failure to learn from military operations. Murray noted that, "military institutions sometimes prove astonishingly resistant to learning from their experiences."[16] Murray's assertion is not without merit. The hierarchy inherent to the military structure can make it difficult for military historians to be completely objective when analyzing past actions, which can lead to inevitable repetition of past mistakes. After analyzing and comparing the logistical effectiveness of the two recent military conflicts in the Persian Gulf, Gary Trogdon, PhD, concluded that military leaders repeated some of the same mistakes in the two conflicts:

> During the first Gulf war logisticians deployed late in the buildup and were unable to get a management handle on incoming cargo, which led to the "iron mountains" of unidentified supplies—the gray boxes—that littered the desert. During the second Gulf war logisticians again deployed late, but faced the additional challenge of being too few in number because of Defense Department-mandated changes in the war plan that created an unbalanced force in the interests of speed and a small footprint in the field of battle.[17]

Trogdon's conclusion raises two concerns. First, is the U.S. guilty of failing to learn from its mistakes, similar to the Germans' failure to do the same over sixty years ago which manifested itself in catastrophe on the steppes of Russia? Second, is America's desire to achieve quick, decisive victories so consuming that she neglects her combat forces at the expense of her support assets? Speed and lethality are important, but operations in the foreseeable future will be a marathon, not a sprint. If the US is going to sustain her commitment to global operations, she must rethink the long-term importance of sustaining those operations. Failure to do so could be disastrous. Trogdon points out:

> In both wars, but especially the more recent one, logistical requirements were sacrificed for combat power to considerable peril. A longer combat phase in the second Gulf war might well have resulted in major logistical shortcomings with a concomitant risk to soldiers on the ground.[18]

The US cannot afford to dismiss the importance of long-term logistical operations. War may be won in a matter of days or weeks, but the commitment incurred may last much longer and require even greater logistical support. The current situation in the Gulf is a prime example. The Germans did not dismiss the importance of their logistical system supporting the battle of Stalingrad. They were obsessed with it. It was the lifeline to an entire army. The point to be made is that enough attention should be paid to logistical operations that they help rather than harm the overall objectives. It is a lesson Hitler would have done well to learn and one the United States cannot afford to ignore.

---

[1]Andrew Brookes, *Air War Over Russia* (Hersham: Ian Allan Publishing, 2003), 10.

[2]Kurt Zeitzler, "Stalingrad," in W. Kreipe, et al., *The Fatal Decision: Six Decisive Battles of the Second World War from the Viewpoint of the Vanquished,* ed. William Richardson and Seymour Freidin, trans. Constantine Fitzgibbon (New York: William Sloane Associates, 1956), 189. Professor Suchenwirth says that, "Hitler, in his desperate eagerness to prevent the Russians from recapturing any sector of the front for fear that he might lose what territory he had already won, ordered his troops to hold out until it was too late for them to escape encirclement. And once a force was encircled, he invariably ordered defensive operations to the last men rather than an immediate attempt to break out." In his editor's note at the bottom of the page, he says, "This was not alone result of Stalingrad, but also a repetition of what Hitler had considered to be his very successful strategy during the winter of 1941-1942 [referring to the Soviet winter offensive that

precipitated the airlifts for Demyansk and Kholm].” Richard Suchenwirth, *Historical Turning Points in the German Air Force War Effort,* USAF Historical Studies, No. 189 (USAF Historical Division, Aerospace Studies Institute, Air University, 1968), 107.

[3]Suchenwirth, 106.

[4]Zeitzler, 189.

[5]Suchenwirth, 106.

[6]Robert S. Dudney, “The Mobility Edge,” *Air Force Magazine* 89, no. 8 (December 2003): 2. Dudney says that, “In March, the first month of the war, 94 percent of all C-5s and 91 percent of all C-17s were committed to worldwide operations. Despite such high utilization rates, there simply wasn’t sufficient lift for the war and the needs of other theaters. . . . There is little doubt the transportation system could not have handled another major crisis and smaller demands in Afghanistan, Bosnia, and Kosovo.” The C-17 is replacing the C-141. While the C-17 can carry almost twice the tonnage of the C-141, the exchange is a 1:2 ratio. Although total airlift tonnage will be about the same, strategic airlift flexibility and U.S. ability to support a lighter, more expeditionary force will be degraded. One C-17 can’t be in two places at once. For more detailed information on the strategic lift deficiencies as a result of the C-17 replacing the C-141, as well as the impact of the C-5’s poor reliability rate, see John A. Tirpak, “A Clamor for Airlift,” *Air Force Magazine* 83, no. 12 (December 2000): 24-30.

[7]Frederic J. Brown, “America’s Army: Expeditionary and Enduring Foreign and Domestic, *Military Review* 83, no. 6 (November-December 2003): 69-77.

[8]Frederick Kagan, “An Army of Lots More Than One,” *The Weekly Standard*, 7 July 2003, [article on-line]; available from www.weeklystandard.com/Utilities/ printer_preview.asp?Article=2849&R=77A627BBD; Internet; accessed 4 Apr 2004. In a separate article, Lt. General John M. Riggs, director of the Objective Force Task Force charged with developing the future Army stated, “I have been in the Army 39 years, and I’ve never seen the Army as stretched in that 39 years as I have today.” This same article also quotes retired Army Gen. Barry McCaffrey, “The Army is going over a cliff by this fall. In my judgment we need 80,000 more troops.” Tom Bowman, “3-Star General Says Army is Too Small To Do Its Job,” *Baltimore Sun* (Baltimore), 21 January 2004.

[9]Robert D. Kaplan, “Supremacy by Stealth,” *The Atlantic Monthly* 292, no. 1 (July/August 2003): 66-80.

[10]Kagan; Bowman.

[11]Williamson Murray, “Clausewitz Out, Computer In: Military Culture and Technological Hubris,” *The National Interest*, 1 June 1997, [journal on-line]; available from http://www. clausewitz.com/CWZHOME/Clause&Computers.html; Internet; accessed 22 January 2004.

[12]Ibid.

[13]This information is based upon numerous discussions with Army officers while attending the Command and General Staff College at Fort Leavenworth, KS during the 2003-04 academic year. CAS3 is also located at Fort Leavenworth.

[14]Numerous Army officers attending the CGSC class of 2004 left the course as much as two months before graduation to meet demands in the field.

[15]Kagan.

[16]Murray continues by saying, "And as difficult as they are to learn in combat, how much harder must it be to learn the lessons of war in peace, absent the harsh, unpredictable, and unforgiving world of death and destruction. Consequently, it is doubly important that in peacetime military professionals work hard to frame the right kind of questions and to generate realistic assumptions," "Clausewitz Out, Computer In."

[17]For a comprehensive comparison of the successes and failures of both Gulf wars, see Gary A. Trogdon, "Logistics in the Desert, December 2003," U. S. Army Center of Military History, Fort McNair, Washington, D.C., 20; for another discussion on the supply problems in the Iraq war, see David Wood, "Military Acknowledges Massive Supply Problems in Iraq War," *Newhouse News Service,* 22 January 2004 [article on-line]; available from http://www.newhouse.com/archive/ wood012204.html; Internet; accessed 24 January 2004.

[18]Trogdon, 20.

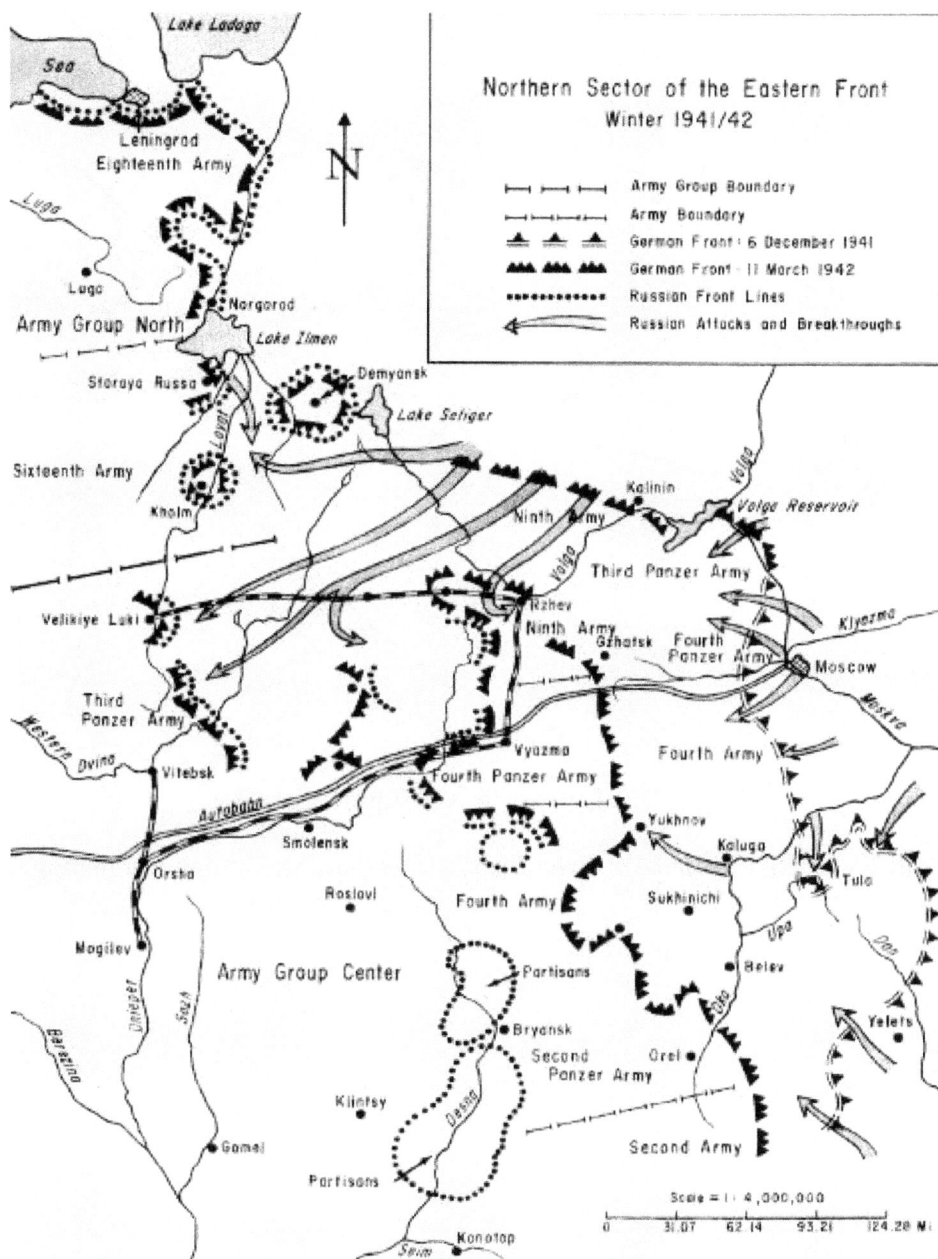

Figure 1.   The Demyansk and Kholm Pockets

Source: Fritz Morzik, *German Air Force Airlift Operations,* USAF Historical Studies, No. 176 (USAF Historical Division, Research Studies Institute, Air University, 1961), 139.

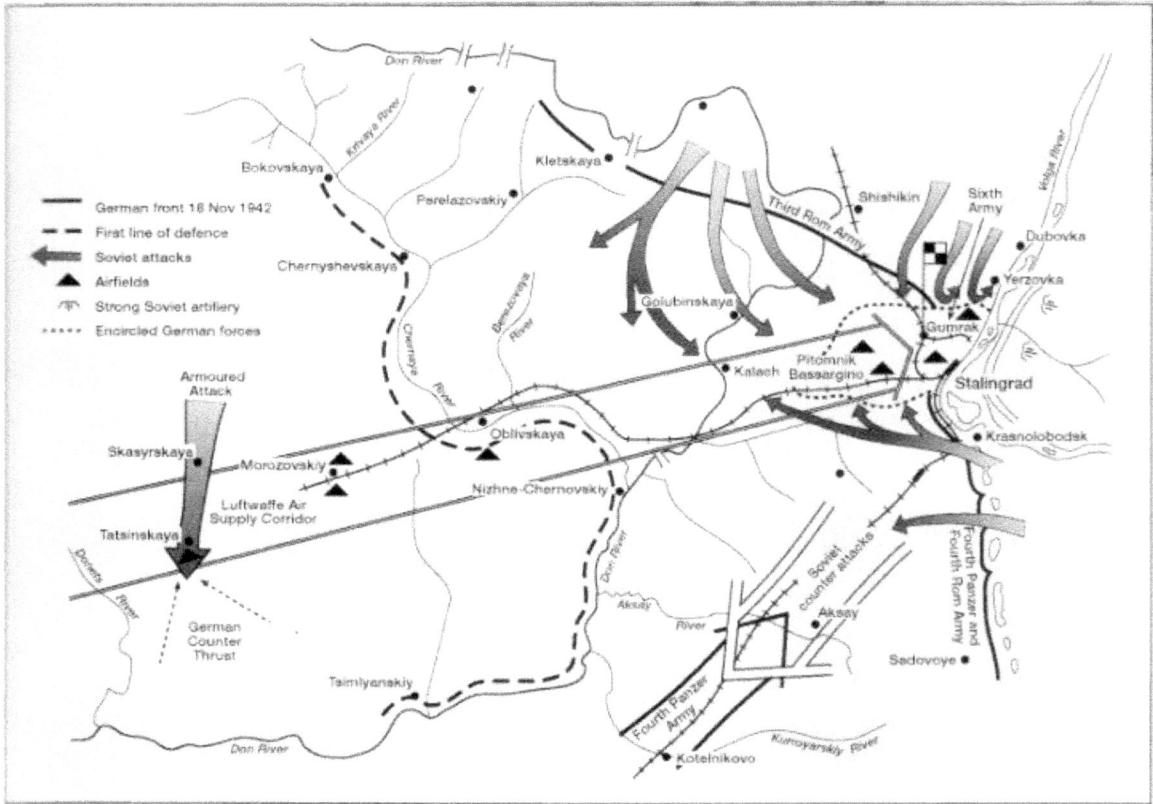

Figure 2.   Luftwaffe Air Supply Corridor

Source: Andrew Brookes, *Air War Over Russia* (Hersham: Ian Allan Publishing, 2003), 99.

SELECTED BIBLIOGRAPHY

Books

Axel, Albert. *Russia's Heroes.* New York: Carroll and Graf Publisher, Inc., 2001.

Battle for Stalingrad: The 1943 Soviet General Staff Study. Ed. Louis C. Rotundo. London: Pergamon-Brassey's International Defense Publisher's, Inc., 1989.

Bekker, Cajus. Translated by Frank Ziegler. *The Luftwaffe War Diaries.* New York: Da Capo Press, 1964.

British Air Ministry. *The Rise and Fall of the German Air Force, 1933-1945.* New York: St. Martin's Press, 1983.

Brookes, Andrew. *Air War Over Russia.* Hersham: Ian Allan Publishing, 2003.

Chuikov, Vasili. *The Battle For Stalingrad.* Translated by Harold Silver. With an introduction by Hanson W. Baldwin. New York: Holt, Rinehart and Winston, 1964.

Cooper, Matthew. *The German Air Force, 1933-1945: An Anatomy of Failure.* New York: Jane's, 1981.

Corum, James S. and Richard R. Muller. *The Luftwaffe's Way of War: German Air Force Doctrine, 1911-1945.* Baltimore: The Nautical and Aviation Publishing Company of America, 1998.

Craig, William. *Enemy at the Gates: The Battle for Stalingrad.* New York: Reader's Digest Press, 1973.

Deichmann, Paul. Edited by Dr. Alfred Price. *Spearhead for Blitzkrieg: Luftwaffe Operations in Support of the Army, 1939-1945.* London, 1996.

Erickson, John. *The Road to Stalingrad: Stalin's War with Germany.* Vol. 1. London: Harper and Row, Publishers, 1975.

Faber, Harold, ed. *Luftwaffe: A History.* New York: Quadrangle, 1977.

Galland, Adolf. *The First and The Last: The Rise and Fall of the German Fighter Forces, 1938-1945.* Translated by Mervyn Savill. New York: Henry Holt and Company, 1954.

Glantz, David M. *A History of Soviet Airborne Forces.* London: Frank Cass, 1994.

Hayward, Joel S. A. *Stopped at Stalingrad: The Luftwaffe and Hitler's Defeat in the East, 1942-1943.* Lawrence, Kansas: University Press of Kansas, 1998.

Irving, David. *The Rise and Fall of the Luftwaffe: The Life of Luftwaffe Marshal Erhard Milch*. Boston: Little, Brown and Company, 1973.

Jukes, Geoffrey. *Stalingrad: The Turning Point*. With an introduction by Captain Sir Basil Liddell Hart. New York: Ballantine Books Inc., 1972.

Kershaw, Ian. *Hitler 1936-45: Nemesis*. New York: W. W. Norton and Company, 2000.

Lee, Asher. *The German Air Force*. New York: Harper and Brothers Publishers, 1946.

Lewis, S. J. *"The Battle of Stalingrad."* In *Block by Block: The Challenge of Urban Operations*. Edited by William G. Robertson and Lawrence A. Yates, 29-58. Fort Leavenworth: U.S. Army Command and General Staff College Press, 2003.

Manstein, Erich von. *Lost Victories*. Novato, Ca.: Presidio Press, 1994.

Murray, Williamson. "Strategic Bombing: The British, American, and German experiences." In *Military Innovation in the Interwar Period*. Edited by Williamson Murray and Allan R. Millet. Cambridge: Cambridge University Press, 1998.

_____. *Strategy for Defeat: The Luftwaffe, 1933-1945*. Maxwell Air Force Base, AL: Air University Press, 1983.

_____. "The World in Conflict." In *The Cambridge Illustrated History of Warfare: The Triumph of the West*. Edited by Geoffrey Parker. Cambridge: Cambridge University Press, 2000.

Nielsen, Andreas. *The German Air Force General Staff*. New York: Arno Press, 1968.

Schneider, Franz and Charles Gullans, trans., *Last Letters from Stalingrad*. With an introduction by S. L. A. Marshall. Westport: Greenwood Press, [1962].

Schröter, Heinz. *Stalingrad*. Translated by Constantine Fitzgibbon. New York: E. P. Dutton and Company, Inc., 1958.

Speer, Albert. *Inside the Third Reich*. Translated by Richard and Clara Winston. New York: The Macmillan Company, 1970.

_____. *Spandau: The Secret Diaries*. Translated by Richard and Clara Winston. New York: Macmillan Publishing Company, 1976.

Stalingrad: An Eye-witness Account. New York: Hutchinson and Company, [1945].

Trevor-Roper, Hugh, ed. *Hitler's Table Talk 1941-44: His Private Conversations*. London: Weidenfeld and Nicolson, 1973.

Trevor-Roper, Hugh, ed. *Hitler's War Directives, 1939-1945*. London: Sidgwick and Jackson, 1965.

Westphal, Siegfried. "Between the Acts." In *The Fatal Decisions: Six Decisive Battles of the Second World War From the Viewpoint of the Vanquished*. Edited by William Richardson and Seymour Freidin. Translated by Constantine Fitzgibbon. New York: William Sloane Associates, Inc., 1956, 190-196.

Zeitzler, Kurt "Stalingrad." In *The Fatal Decisions: Six Decisive Battles of the Second World War From the Viewpoint of the Vanquished*. Edited by William Richardson and Seymour Freidin. Translated by Constantine Fitzgibbon. New York: William Sloane Associates, Inc., 1956, 132-189.

Ziemke, Earl and M. E. Bauer. *Moscow to Stalingrad: Decision in the East*. Washington D.C.: Center of Military History, United States Army, 1987.

_____. *Stalingrad to Berlin: The German Defeat in the East*. Washington, D.C.: Center of Military History, United States Army, 1987.

U.S. Government Sponsored Documents

Primary

U. S. Army. Command and General Staff College. *Headquarters Air P/W Interrogation Detachment Military Intelligence Service: Hermann Goering*. 1 June 1945. Combined Arms Research Library, Fort Leavenworth, KS. File N-10007-3.

_____. *Interrogation of Reich Marshall Hermann Goering. 10 May 1945, 1700 to 1900 hours*. Combined Arms Research Library, Fort Leavenworth, KS. File N-9618.

Secondary

Corum, James S. "The Development of Strategic Air War Concepts in Interwar Germany, 1919-1939," taken from Air Command and Staff College. *Distance Learning Version 3.0: Military Studies*. Montgomery, AL: Air Command and Staff College, 2000.

Deichmann, Paul. *German Air Force Operations in Support of the Army*. USAF Historical Studies, No. 163. Montgomery, AL: USAF Historical Division, Research Studies Institute, Air University, 1962.

Emme, Eugene. *Hitler's Blitzbomber*. Maxwell Air Force Base, AL: Documentary Research Division, Research Studies Institute, Air University, 1951.

House, Jonathan M. *Toward Combined Arms Warfare: A Survey of 20th-Century Tactics, Doctrine, and Organization.* Fort Leavenworth: U.S. Army Command and General Staff College, 1984.

Howell, Edgar. *The Soviet Partisan Movement, 1941-1944,* Department of the Army Pamphlet 20-244. Washington: U.S. Government Printing Office, 1956.

Morzik, Fritz. *German Air Force Airlift Operations.* USAF Historical Studies, No. 167. Montgomery, AL: USAF Historical Division, Aerospace Studies Institute, Air University, 1961.

Plocher, Herman. *The German Air Force versus Russia, 1942.* USAF Historical Studies, No. 154. Montgomery, AL: USAF Historical Division, Aerospace Studies Institute, Air University, 1966.

_____. *The German Air Force versus Russia, 1943.* USAF Historical Studies, No. 155. Montgomery, AL: USAF Historical Division, Aerospace Studies Institute, Air University, 1967

Spiller, Roger. *Sharp Corners: Urban Operations at Century's End.* Fort Leavenworth, KS: Combat Studies Institute, 2001.

Trogdon, Gary A. "Logistics in the Desert, December 2003." U.S. Army Center of Military History, Fort McNair, Washington, DC.

"The United States Strategic Bombing Survey: Summary Report (European War)," ed. US Army Command and General Staff College. *H100 Transformation in the Shadow of Global Conflict.* Fort Leavenworth: U. S. Army Command and General Staff College, 2003.

Vaughn, David K. and James H. Donoho. "From Stalingrad to Khe Sanh: Factors in the Successful Use of Tactical Airlift to Support Isolated Land Battle Areas." *Air and Space Power Chronicles.* Maxwell Air Force Base Database on-line. Available from http:// www.airpower.maxwell.af.mil/airchronicles/cc/vaughan.html. Internet. Accessed 15 August 2003.

Dissertations and Theses

Muller, Richard. "The German Air Force and the campaign against the Soviet Union, 1941-1945." Ph.D. Diss., The Ohio State University, 1990.

Thyssen, Mike. "A Desperate Struggle to Save A Condemned Army--A Critical Review of the Stalingrad Airlift." Research Paper, Air Command and Staff College, 1997.

Periodicals and Newspaper Articles

Brown, Frederic J. "America's Army: Expeditionary and Enduring Foreign and Domestic. *Military Review* 83, no. 6 (November-December 2003): 69-77.

Dudney, Robert S. "The Mobility Edge." *Air Force Magazine* 86, no. 8 (August 2003): 2.

Gosztony, Peter I. "22 June 1941." *Military Review* 51, no. 6 (June 1971): 47-51.

Kaplan, Robert D. "Supremacy by Stealth." *The Atlantic Monthly* 292, no. 1 (July/August 2003): 66-80.

Kagan, Frederick. "The Evacuation of Soviet Industry in the Wake of 'Barbarossa': A Key to the Soviet Victory." *The Journal of Slavic Military Studies* 8, no. 2 (June 1995): 387-414.

Murray, Williamson. "Clausewitz Out, Computer In: Military Culture and Technological Hubris." *The National Interest*, 1 June 1997. Journal on-line. Available from http://www. clausewitz.com/CWZHOME/Clause&Computers.html. Internet. Accessed 22 January 2004.

Pickert, Wolfgang. "The Stalingrad Airlift: An Eyewitness Commentary." *Aerospace Historian,* (December 1971): 183-185.

Potter, Edwin J. "Prelude to Barbarossa." *Military Review* 58, no. 9 (September 1968): 56-64.

Sas, Anthony. "Invasion of Russia." *Military Review* 51, no. 6 (June 1971): 38-46.

Tirpak, John A. "A Clamor for Airlift." *Air Force Magazine* 83, no. 12 (December 2000): 24-30.

Wood, David. "Military Acknowledges Massive Supply Problems in Iraq War." *Newhouse News Service,* 22 January 2004. Article on-line. Available from http://www.newhouse.com/archive/ wood012204.html. Internet. Accessed 24 January 2004.

INITIAL DISTRIBUTION LIST

Combined Arms Research Library
U.S. Army Command and General Staff College
250 Gibbon Ave.
Fort Leavenworth, KS 66027-2314

Defense Technical Information Center/OCA
825 John J. Kingman Rd., Suite 944
Fort Belvoir, VA 22060-6218

LTC John A. Suprin
CSI
USACGSC
1 Reynolds Ave.
Fort Leavenworth, KS 66027-1352

Dr Samuel J. Lewis
CSI
USACGSC
1 Reynolds Ave.
Fort Leavenworth, KS 66027-1352

Col William J. Heinen
AFELM
USACGSC
1 Reynolds Ave.
Fort Leavenworth, KS 66027-1352

CERTIFICATION FOR MMAS DISTRIBUTION STATEMENT

1. Certification Date: 18 June 2004

2. Thesis Author: Major Willard B. Akins II

3. Thesis Title: The Ghosts of Stalingrad

4. Thesis Committee Members: _____

   Signatures: _____

   _____

5. Distribution Statement: See distribution statements A-X on reverse, then circle appropriate distribution statement letter code below:

(A) B C D E F X   SEE EXPLANATION OF CODES ON REVERSE

If your thesis does not fit into any of the above categories or is classified, you must coordinate with the classified section at CARL.

6. Justification: Justification is required for any distribution other than described in Distribution Statement A. All or part of a thesis may justify distribution limitation. See limitation justification statements 1-10 on reverse, then list, below, the statement(s) that applies (apply) to your thesis and corresponding chapters/sections and pages. Follow sample format shown below:

EXAMPLE

Limitation Justification Statement	/	Chapter/Section	/	Page(s)
Direct Military Support (10)	/	Chapter 3	/	12
Critical Technology (3)	/	Section 4	/	31
Administrative Operational Use (7)	/	Chapter 2	/	13-32

Fill in limitation justification for your thesis below:

Limitation Justification Statement	/	Chapter/Section	/	Page(s)
_____	/	_____	/	_____
_____	/	_____	/	_____
_____	/	_____	/	_____
_____	/	_____	/	_____
_____	/	_____	/	_____

7. MMAS Thesis Author's Signature: _____

STATEMENT A: Approved for public release; distribution is unlimited. (Documents with this statement may be made available or sold to the general public and foreign nationals).

STATEMENT B: Distribution authorized to U.S. Government agencies only (insert reason and date ON REVERSE OF THIS FORM). Currently used reasons for imposing this statement include the following:

    1. Foreign Government Information. Protection of foreign information.

    2. Proprietary Information. Protection of proprietary information not owned by the U.S. Government.

    3. Critical Technology. Protection and control of critical technology including technical data with potential military application.

    4. Test and Evaluation. Protection of test and evaluation of commercial production or military hardware.

    5. Contractor Performance Evaluation. Protection of information involving contractor performance evaluation.

    6. Premature Dissemination. Protection of information involving systems or hardware from premature dissemination.

    7. Administrative/Operational Use. Protection of information restricted to official use or for administrative or operational purposes.

    8. Software Documentation. Protection of software documentation - release only in accordance with the provisions of DoD Instruction 7930.2.

    9. Specific Authority. Protection of information required by a specific authority.

    10. Direct Military Support. To protect export-controlled technical data of such military significance that release for purposes other than direct support of DoD-approved activities may jeopardize a U.S. military advantage.

STATEMENT C: Distribution authorized to U.S. Government agencies and their contractors: (REASON AND DATE). Currently most used reasons are 1, 3, 7, 8, and 9 above.

STATEMENT D: Distribution authorized to DoD and U.S. DoD contractors only; (REASON AND DATE). Currently most reasons are 1, 3, 7, 8, and 9 above.

STATEMENT E: Distribution authorized to DoD only; (REASON AND DATE). Currently most used reasons are 1, 2, 3, 4, 5, 6, 7, 8, 9, and 10.

STATEMENT F: Further dissemination only as directed by (controlling DoD office and date), or higher DoD authority. Used when the DoD originator determines that information is subject to special dissemination limitation specified by paragraph 4-505, DoD 5200.1-R.

STATEMENT X: Distribution authorized to U.S. Government agencies and private individuals of enterprises eligible to obtain export-controlled technical data in accordance with DoD Directive 5230.25; (date). Controlling DoD office is (insert).